LOGIC magazine

10

security
spring 2020

LOGIC

ISSUE 10: Security

EDITORIAL

Ben Tarnoff, Jen Kagan, Alex Blasdel, Moira Weigel

CREATIVE

Xiaowei R. Wang, Celine Nguyen

PRODUCTION

Jim Fingal, Christa Hartsock

/*

Cover & interstitials: Xiaowei R. Wang

Images: p. 42, 43 Google Patents.

ISBN (print)	978-1-952550-00-3
ISBN (digital)	978-1-952550-01-0
ISSN	2573-4504
BLOOPS (call)	419-452-4857

ELECTRONIC TELEGRAM

editors@logicmag.io

INTERNET

https://logicmag.io

Logic Magazine is published by the Logic Foundation, a California nonprofit public benefit corporation organized for the specific purpose of promoting education about technology.

*/

security

LOGIC MAGAZINE

Editorial

Chatlogs

ISSUE 10
SECURITY

Features

Patches

Assets

Contributors

American Artist is an interdisciplinary artist whose work considers Black labor and visibility within networked life.

Mitch Anzuoni is an editor and writer based in New York City. They help run Inpatient Press, an independent press dedicated to esoteric and experimental works.

Gabriella Coleman holds the Wolfe Chair in Scientific and Technological Literacy at McGill University.

Veena Dubal is an Associate Professor of Law at the University of California, Hastings. Her research focuses on the intersection of law, technology, and precarious work.

Matt Goerzen studies sociotechnical security issues at the Data & Society Research Institute.

Darius Kazemi is an internet artist, software developer, and moderator of Friend Camp.

Joyce S. Lee is a UX researcher based in Oakland, California.

Darren Loucaides is a writer based in Barcelona and London.

Alison Macrina is a librarian, internet activist, and founder and executive director of the Library Freedom Project.

Tawana "Honeycomb" Petty is a mother, social justice organizer, youth advocate, poet, and author. She directs the Data Justice program for Detroit Community Technology Project and co-leads the Our Data Bodies project.

Astra Taylor's latest book is *Democracy May Not Exist, but We'll Miss It When It's Gone.*

Rian Wanstreet is a PhD candidate at the University of Washington and an Open Science Fellow with Mozilla, researching the ramifications of new technologies in the agricultural sector.

Emma Weil is a research assistant at Upturn in Washington, DC.

From the Editors

Meditations in an Emergency

1/

As we close this issue, COVID-19 case numbers are surging across the European Union, and if they are not yet as high in North America, it seems to be mostly for a lack of tests. Oil prices are plunging, the Dow Jones is plunging, and Ted Cruz is in voluntary self-quarantine. New York State prisoners are making hand sanitizer for $0.65 per hour. Passengers are disembarking from the Diamond Princess into the Port of Oakland.

For months, prominent figures in the tech industry have been warning that it will get worse before it gets better.

In February, *Recode* reported that the venture capital firm Andreesen Horowitz was already on high alert, canceling employee travel to China and banning handshakes in the office. In early March, Sequoia Capital wrote a memo warning partners that coronavirus would be the “black swan” of 2020; they urged startups to prepare for leaner times.

On Twitter, one "founder coach and angel investor" surmised that "people in Silicon Valley were taking #COVID19 more seriously than other parts of the US including NYC" because "SV understands the power of network effects and exponential growth." "Congrats Silicon Valley you took a math class," a tech journalist turned venture capitalist swiped back.

Everyone was on edge.

The exponential growth in tweets about exponential growth was easy to make fun of, combining, as it did, several stereotypes. The prepper engineer stocking Soylent. The billionaire in his New Zealand bunker. The thought leader inclined to consider himself an expert and therefore obliged to opine on everything, since everything is, or can be made into a source of, data. And yet. There was something to what the Founder Coach and Angel Investor said.

To tell a good horror story, it is not enough to imagine some freak occurrence. You have to find what is frightening about what is ordinary—or, more precisely, ordinarily desired. A large home with a pool, backyard, and picture windows. A movie star husband. A beach vacation. A hot shower after a long drive. The very thing you wanted must turn deadly. Otherwise, you could just walk away.

The pacing of the unfolding pandemic has more in common with a conspiracy theory than with a slasher flick. Its mode is twitchy simultaneity rather than the drumbeat of a killer closing in. *Now our victim is in Wuhan. Now in Bergamo. Now New York. The villain is invisible and everywhere.*

But as the coronavirus spread all over the world, the tech industry watched the properties that it usually thinks of as strengths turn into vulnerabilities. Disruption was rampant. The same transportation networks, and supply chains, that had turned second-tier cities into "furnaces" fueling global just-in-time production suddenly became vectors for infection. The same networks that could be used to track cases, and disseminate information, were promoting dangerous lies. What we are watching, at time of writing, from home, is hockey-stick growth, but for sickness and death.

2/

Viral, we call content that spreads quickly by means of preexisting bodies and behaviors. A virus takes our tendencies to make new cells, or touch one another, or share a laugh, and turns them to its own sole purpose: self-propagation. What makes something catch on is a subject of much speculation. What is clear is that some dangers cannot be confronted without considering the ways that the entire system they imperil works.

This issue will explore forms of security and insecurity that arise as digital technologies enter new realms of existence — and how stubbornly these two terms intertwine.

In one sense, the dynamics that it explores are not new.

The rise of capitalism in the fifteenth and sixteenth centuries provided some insurance against the whims of, say, the weather. Labor markets promised to free workers from physical coercion: rather than toiling on the particular piece of land where you happened to have been born, or where some lord put a sword to your head and told you, *Toil!,* you could choose where to go work where you wanted, for a wage.

And yet, in order to get most people to work for wages at all, those in power had to produce widespread insecurity. By turning life's necessities into commodities that had to be bought and sold on a market, they made it impossible for most people to survive any other way. Digital technologies have only intensified this dynamic. To access information — a matter of survival in a pandemic — we rely on the platforms. If you want to get information about what the pandemic means for your kid's school, you will likely have to join Facebook.

The current crisis has brought us back to these founding facts of capitalism as a world system. The market mediates our every move, and binds us, through countless threads of interdependence, to everyone else. Nothing like shortages of toilet paper to highlight how few of us would be capable of single-handedly reproducing the conditions of our lives.

At the same time, technologies designed to make some people and property more secure create new vulnerabilities. Machine learning algorithms used to plant, water, and harvest miles of crops at the optimal times can be hacked; a software bug or glitch can now cause a famine. Wireless technologies allow the lender that has loaned you the money to buy the car you need to drive to work to remotely disable that car if you should fall behind on monthly payments.

Securing one man's loan can endanger another man's livelihood or life—or both, should the remote disabling, for instance, happen while you drive to work on the freeway. And so, when we talk about security, we always have to ask: Secure *from what?* Secure *for whom?*

3/

You can use a baseball bat to play a game or break a window; you can set a table, or kill your hostess, in the parlor, with a candlestick.

Any tool can be abused, and if we have learned anything in the years since Gamergate and the 2016 presidential election, it is that some systems are most dangerous when they are being used as they were designed to be. The same features that make it possible for social networks to produce "cognitive surplus" and new social movements also produce misinformation and harassment.

The authors in this issue explore how the very same products and design features that are supposed to make some secure make others less so. Code designed to let you prove that *You are not a robot* will not make you safer if you are visually impaired and need your phone to play the code aloud every time you log in. Privately owned surveillance cameras, whose footage is constantly scanned by proprietary facial recognition software, are unlikely to make all the residents of a city safer. As one activist interviewed in these pages puts it: *Feeling watched is not the same as feeling seen.*

4/

It is taking too long to write this. We can hardly keep up with our push notifications.

As we close this issue, deaths in Italy surpassed deaths in China; the number of confirmed cases in the United States passed 30,000, still with a dearth of tests; California sheltered in place; at least hundreds of thousands of Americans are losing their jobs, applying en masse to a safety net that is not ready for the load; and at least two senators were found to have sold millions of dollars in stock after attending a closed door coronavirus briefing, and then lying to the public about the severity of the coming crisis. It is probably too much to hope that they do not sleep soundly. Their colleague Rand Paul became the first senator to test positive for coronavirus; he is not likely to be the last.

Meanwhile, powerful tech firms are finding ways to play the situation to their advantage. The alerts continue: Amazon has announced that they will hire 100,000 more workers to meet skyrocketing demand for "contactless delivery," as their competitor businesses stand empty or close down. Palantir and Clearview AI are making bids to sell software to surveil infected people and their networks to the state; Facebook and Google are offering to use the track-location function in their apps to do the same.

In China, the pandemic has intensified, and legitimated, expansive government control. The tools that are making it possible for millions of Americans to learn and work remotely are also creating new opportunities for both public and private entities to do the same. School-issued computers come with spyware installed; Zoom video conferencing software allows the boss to track "attendee attention" or drop in at any time.

5/

What would it take to turn this emergency inside out?

An emergency produces opportunities for solidarity as well as selfishness. A pandemic does this in particular. *To the idiots buying five bottles of soap*, one viral tweet admonished, *it doesn't matter how many times you wash your hands if your neighbors cannot wash theirs*. Our present situation highlights, with particular starkness, that nobody can be secure alone.

Ancient philosophers argued that care or worry, *cura*, was the defining feature of human life. The state of being *sine cura*—"without" or free from it—was, at best, fleeting. On the other hand, care or worry is also the origin of *curiosity*.

This issue proposes that the concept of security can still hold the same ambivalence and cautious hopefulness now as it did then. We must care for our neighbors not in spite of but because of our common vulnerabilities. (In Toledo, Ohio dozens of volunteers have begun sewing fabric masks for local doctors and nurses already running out of them.) As the social contagion of panic spreads, we hope this idea catches on, too: that the very same fears pushing millions of people into isolation can also draw us outward, binding us more firmly to the world and one another.

019

The Insecurity Machine

by Astra Taylor

Capitalism generates security for the few and insecurity for the many. Digital technologies are making it worse.

Long before COVID-19 swept the globe, insecurity was already everywhere. Countless people faced housing, health, and food insecurity. Environmental insecurity was rising as changing weather patterns put communities at risk of fires and flooding. Prior to the advent of "social distancing," we hid behind doors, locks, gates, and border walls, afraid of public space and one another. Online, we fretted over information security, devising passwords to access passwords, fearful we might be hacked or exposed. We were insecure at our jobs, in our homes, in our relationships, and on social media. We felt insecure about our very selves.

Given the ubiquity of insecurity, it may seem surprising that, only a few centuries ago, the word didn't even exist. A uniquely modern concept, insecurity first appeared in the seventeenth century. "Rather than being understood as an unalterable truth intrinsic to the human condition, 'insecurity' needs to be understood as the product of very specific historical circumstances," the political theorist Mark Neocleous observes. Those particular historical circumstances were the rise of capitalism.

Consider the response to the coronavirus. In the United States, the devastation unleashed by this novel and dangerous pathogen has as much to do with economics as epidemiology. When whole sectors of the economy shut down to comply with orders to shelter in place, unemployment soared. Instead of paying workers to stay home to slow down the disease—a sensible, life-saving policy pursued in various ways by some wealthy countries, including Denmark—American officials handed trillions of dollars of public money to the world's biggest corporations. For the CEOs of these companies, the outbreak is less a crisis than an opportunity. Not only will they receive a staggering no-strings-attached government handout, they'll also get a more pliable and profitable workforce. Experts estimate that unemployment rates could hit 30 percent, higher than during the Great Depression. When this pandemic passes, millions of people will be even more insecure and exploitable than they were at the outset, and that will not be an accident.

Capitalism is an insecurity machine, though we rarely think of it as such. Alongside profits, commodities, and inequality, insecurity is a fundamental output of the system. Neither an incidental byproduct nor a secondary consequence of the concentration of wealth, it is one of capitalism's essential and enabling creations. "The bourgeoisie cannot exist without constantly revolutionizing the instruments of production, and thereby the relations of production, and with them the whole relations of society," Marx and Engels wrote in *The Communist Manifesto*, back when the most advanced machines were weaving cloth and harnessing steam. "Constant revolutionizing of production, uninterrupted disturbance of all social conditions, everlasting uncertainty and agitation distinguish the bourgeois epoch from all earlier ones." The beneficiaries of this arrangement dubbed it "creative destruction" before rebranding it as "disruption."

Our economic apparatus, in other words, destabilizes by design: market forces capsize communities and disintegrate old ways of life. Too often, however, we emphasize dramatic shifts, underscoring "great transformations" and systemic crises over comparatively quotidian developments—a temptation we must resist as we face the dual calamity of a global pandemic and an economic downturn. Engineered in order to facilitate exploitation and undermine solidarity, the production of insecurity is a daily phenomenon, its operations so commonplace as to appear banal. It is both physical and psychological: people endure inadequate housing, wages, and healthcare while our culture encourages self-blame and shame for financial hardship, relentlessly exploiting our fears and vulnerabilities. (No advertisement will ever tell us we're okay and that it is the world that needs changing.)

Of course, people have always lived precarious lives. Long before the industrial revolution, let alone the digital one, human existence was neither easy nor assured. (This partly explains why "security," unlike "insecurity," is an ancient concept and aspirational ideal: etymologically, security comes from the Latin *securitas*, meaning freedom from worry, *sine cura*, or without care.) In a similar fashion, commerce precedes capitalism; people have long engaged in the propensity to truck, barter, and trade. Capitalism emerges when the possibility of commerce becomes the necessity of competitive production. For that to happen—for market *opportunities* to become market *imperatives*—mass insecurity must be imposed and maintained.

Digital technologies provide new channels through which this process can unfold. Social media elevates paid advertising and sensational content, spreading misinformation and confusion, increasing epistemological insecurity. Data brokers create intimate profiles so that we might be better targeted—segmenting us into categories that include "rural and barely making it," "probably bipolar," and "gullible elderly"—while companies invest millions into "affect recognition" so they can figure out when we are most persuadable, increasing psychological insecurity. Opaque systems of information collection and predictive analytics facilitate new forms of discrimination and redlining, marking certain populations as criminal threats or directing them into subprime financial services, predatory mortgages, and exploitative rental markets, increasing housing insecurity. Employers monitor and control employees remotely, refusing to offer decent wages and benefits or provide consistent scheduling, increasing job insecurity.

Not everyone is made equally insecure by these tools, of course. That is precisely the point. The stability of ownership and investment for some necessarily depends on the destabilization and dispossession of others—and the struggle over housing and labor have always been the epicenters of this conflict. The spaces where we live and where we work are capitalism's main battlegrounds, and the rise of networked digital technologies have given capital more powerful weapons with which to conquer them—weapons we can be assured will be put to use as we enter a phase of coronavirus-induced uncertainty and volatility.

Whose Streets?

In the summer of 2019, the residents of Atlantic Plaza Towers, a 718-unit apartment building in Brownsville, Brooklyn, got word that their landlord had plans to install a security system equipped with facial recognition. The devices would control entry to the twin high-rises and monitor all common areas.

The grounds were already covered in cameras. So who or what, exactly, would this new technology be making more secure, the tenants wondered. Contrary to the landlord's insistence that the devices were a "cool upgrade" that would keep keys out of the hands of the "wrong people," residents saw them as an unwelcome and unwanted intrusion—and one inextricably linked to a long and troubling history of racism, policing, and gentrification.

> *"The stability of ownership and investment for some necessarily depends on the destabilization and dispossession of others."*

The mostly Black and female tenants were alarmed to learn of studies showing that facial recognition often perpetuates pre-existing bias, with software most accurately able to assess men with white skin. They also worried about the collection of sensitive biometric data. "I'm afraid of it being shared with third-party agencies. I'm afraid of it being shared with the police. I'm afraid of it being shared with anyone—advertising companies, just everyone. It's just very sensitive information that I feel our landlord should not have," a young woman named Tranae Moran told the *Guardian*. Another resident, who had called the complex home for fifty-one years, was more blunt: "We do not want to be tagged like animals. We are not animals. We should be able to freely come in and out of our development without you tracking every movement."

One hundred thirty-four tenants filed a formal complaint. By banding together and partnering with lawyers, they thwarted the landlord's plan and made national news. Their success, however, represents more than a victory for privacy rights or the growing backlash against a particularly problematic technology. (Thanks to the work of activists, the use of facial recognition software by government agencies has been banned in a handful of American cities.) On a deeper level, the tenants identified and resisted one of capitalism's central dynamics: the fact that security for some is predicated on the insecurity of others.

While the landlord never said it directly, the protection of an investment, not the community, drove the adoption of the new camera system. Increased "security" was part of a bid to attract new, more affluent tenants, whose arrival would cause rents and property values to rise, threatening long-time residents with displacement. "He doesn't want Spanish. He doesn't want Black. He wants white people to come into the neighborhood," one tenant put it.

What happened in Brownsville is a variation of an old story, albeit with a high-tech twist. Property investments always involve what geographer Brett Story calls "the coercive scaffolding of enclosure and securitization"—a scaffolding that harkens back to capitalism's murky origins in the English countryside. During the long and varied period called the enclosure movement, beginning in the twelfth century, wealthy landlords uprooted the peasantry in order to privatize once communal fields and forests, denying them their customary rights to the commons. A contemporary version of this dynamic now plays out in America's fast-gentrifying cities, where crime is redefined as a threat to the real estate industry's bottom line. Homeless people are the initial target, cleared off the streets by ordinances that outlaw panhandling or simply sitting on the sidewalk. Casting the poor not as residents deserving of rights but as miscreants on the wrong side of the law justifies their exclusion.

Nowhere is this more apparent, Story argues, than modern-day Detroit, where a coalition of real estate investors, including the billionaire founder of Quicken Loans, Dan Gilbert, are pushing forward a "revitalization" effort. As was almost the case in Brownsville, electronic monitoring plays a key role. In Detroit's downtown Chase Tower, a command center contains dozens of computer screens connected to approximately 1,000 different outdoor cameras surrounding Gilbert's properties in seven states, which include over 300 in metro Detroit alone. "The camera program is a collaborative effort that includes most of the big downtown property owners, including General Motors, Ilitch Holdings, and Compuware," Story writes, one that also coordinates closely with local law enforcement agencies. Dangling the prospect of economic growth, real estate moguls are able to redirect public power to private ends: the security guards that Gilbert employs can use force on civilians but are under no legal obligation to read detainees their Miranda rights.

In 1843, a young Marx described security—what he called "the concept of *police*"—as the "supreme concept" of bourgeois society. Fearful of those they have dispossessed, ruling elites have long utilized state violence to safeguard private assets, criminalizing both poverty and protest in the process. In 2016, Detroit's public-private surveillance system was used to track Black Lives Matter protesters. That same year, business owners formed a consortium called Project Green Light, which enables them to stream their surveillance footage directly to city police facilities. "Now we don't have just one billionaire (doing it), we have 500 businesses who pitch in and do their own areas," Detroit's mayor Mike Duggan boasted last spring. Six months later, grassroots opposition successfully blocked the police department from using real-time facial recognition on Green Light's feeds, though the police continue to use the controversial technology in other ways.

Coding the Invisible Hand

Ever since Thomas Hobbes portrayed civilized man trading obedience for protection to escape a perilous "state of nature," security has been central to the liberal political tradition. Government, Adam Smith proclaimed, exists "for the security of property." Similarly, John Locke insisted that the reason men put "themselves under Government is the Preservation of their Property." Yet in Locke's view, not all property deserved to be preserved: he defended the British seizure of Indigenous territory in the Americas. The question, then as now, is *who* and *what* is being secured—and at whose expense.

Today, market logic so suffuses the concept of security that the term literally means property, like the security deposit you make before signing a lease or the "securities" owned by the affluent. It was these ironically-named securities that brought down the global economy in 2008. Traders, using algorithms that coded Black borrowers and homeowners as particularly exploitable, gambled with securitized mortgages boasting inflated ratings. At the same time, the multibillion-dollar "lead generation" industry, which uses digital tools to compile and sell lists of prospective online customers, enabled lenders to identify potential subprime borrowers. This process, experts say, "played a critical, but largely invisible, role" in the mortgage crisis. In the end, nine million families saw their homes foreclosed on, wiping out half the collective wealth of Black families nationwide, further devastating deindustrialized cities like Detroit. With a few strokes of a keyboard, modern bankers caused dispossession on a scale that put the landowners of the original enclosure movement to shame.

The havoc wrought by the mortgage crisis in turn opened space for new algorithmically enabled land grabs. Invitation Homes, a private equity-backed firm, broke new ground by adopting machine learning systems to assess rental acquisitions, buying up huge swaths of property foreclosed during the subprime crisis on the wager that Americans would be willing to rent suburban houses they could no longer afford to own. Today, one-fourth of single-family rentals belong to institutional investors, which have developed streamlined, smartphone-driven systems to manage them. Digital technology mediates all interactions between tenants and the company, from viewings to lease signings to repairs, inaugurating what scholar Desiree Fields calls the age of the "automated landlord." (In March 2020, New York Governor Andrew Cuomo tasked William Mulrow, a senior director at the private equity giant Blackstone Group, with spearheading the state's coronavirus economic recovery; Blackstone held a major stake in Invitation Homes until November 2019, when it sold its last shares for a total of around $7 billion.)

As Fields has documented, the digital dimensions of this high-tech landgrab go much deeper than a shiny interface. An assemblage of platforms and data analytics drive what the National Rental Home Council, a trade association, describes as "property management at scale." First, Invitation Home's proprietary underwriting algorithm determines what properties the company should purchase by considering factors including "neighborhood desirability, proximity to employment centers, transportation corridors, community amenities, construction type, and required ongoing capital needs." Then networked technology allows investors to oversee large portfolios of far-flung units, with information quickly conveyed to capital markets so that additional money can be raised to expand the enterprise, while regular people who want to purchase a place to live are forced to compete with distant cash-rich investors working at digital speed. Hedge funds are happy to let buildings sit empty, waiting for them to appreciate, while locals pay the price.

Meanwhile, old biases persist and compound even when the platform is cutting-edge. Invitation Homes, for example, targets people of color who lack other housing options while charging them sky-high rates to meet Wall Street's outsized expectations. In other instances, opaque systems make discrimination difficult to prove. Automated decision-making enshrines socioeconomic disparities in an invisible, technical process, locking certain populations out or including them on predatory terms. One recent study from UC Berkeley found that, among online mortgage applicants, Black and Latinx borrowers paid over five basis points more in interest than white borrowers with similar financial backgrounds.

> *"Today, employees no longer need to labor in the same physical space to be surveilled, nor is a human being required to do the surveilling."*

Racism is encoded in bad datasets and reinforced by the biases of disproportionately white, male, and privileged engineers—a process scholar Ruha Benjamin calls the "New Jim Code." Recently, the Trump Administration's Department of Housing and Urban Development proposed new rules that would effectively permit automated discrimination in the housing market, allowing algorithms to exclude and segregate on a landlord or mortgage lender's behalf, effectively exempting digital technology from civil rights regulations. "It's going to drive people toward these algorithmic tools, and I think we'll end up in a marketplace where everyone is taking advantage of this loophole," Paul Goodman, a housing justice advocate, told *Dissent*. The powerful may soon be allowed to have computers mark certain populations as "risky" in order to dispossess them, and to do so without risking a lawsuit.

All Watched Over by Machines at the Workplace

New technologies aren't just augmenting capitalism's insecurity-generating tendencies in the spaces that we call home. They are also intensifying those tendencies in the other domain that defines most of our lives: the workplace.

The other night a friend regaled me with stories of working at a Brooklyn café. The place has a vintage and vaguely Parisian aesthetic—decidedly retro and low-tech. There are, of course, regulars, including a medievalist who likes to chat. A few months ago, on a slow day, another barista on duty was exchanging pleasantries with the medievalist when her phone rang: the owner was watching the security camera live-feed from his laptop, and told her to stop being so nice. When I asked my friend how many cameras are installed in the small space, she could identify at least eight, and said there might be more. The charming café is, in fact, a panopticon—the boss can tune in at any time from anywhere, and see from nearly every angle. The workers are always on edge, even when all they want to do is show a bit of kindness to a local eccentric.

As the scholar George Rigakos reminds us in his book *Security/Capital*, employers have been deploying cameras toward similar ends for decades. In the early 1990s, Rigakos worked at a bakery where the staff regularly took home broken and unsaleable loaves. Management had always looked the other way. But weeks before the business was scheduled to be closed, the owners installed security cameras to catch workers in the act. Lifelong employees were summarily fired, losing their retirement benefits. "The security cameras must have saved the company thousands upon thousands in severance and pension dollars," Rigakos recounts.

Today, employees no longer need to labor in the same physical space to be surveilled, nor is a human being required to do the surveilling. Instead, isolated and geographically dispersed, workers can be tracked and controlled remotely, whether they are driving for UPS or making deliveries for DoorDash or transcribing material for Rev. By harnessing digital technology, companies are able to offload more risk onto individuals, whom they categorize as independent contractors to bypass minimum wage laws and other protections. A dwindling number of people are entitled to severance or pension dollars in the first place.

Work, we are often told, is becoming insecure. In reality, insecurity precedes work, or at least its waged variety. While some claim that insecurity is the inevitable consequence of innovation—the result of the fact that, as labor productivity rises, you need fewer workers to produce the same output—the fact is that people had to be *made* insecure, literally severed from their land and livelihoods, for capitalist working conditions to be foisted upon them.

Before the wage-earner could emerge as our society's paradigmatic subject, the persona that we must all embody to survive, the condition of what historian Michael Denning calls "wagelessness" had to be imposed via the process of enclosure, after which peasants could no longer provide for themselves. "Capitalism begins not with the offer of work, but with the imperative to earn a living," Denning writes. Contrary to the myth of liberal laissez-faire, employment relations are anything but natural, spontaneous, or freely chosen.

It wasn't until during the New Deal era that employment became secure, at least for a subset of white men. During the Great Depression, an unlikely assortment of social reformers, radical workers, and "welfare populists" pushed to redefine "security" as a social good guaranteed by the government. "For a long time now people have been saying that perhaps the greatest evil of capitalist industrialism is not its unequal distribution of wealth but the insecurity it brings to the majority of the population," *The New Republic* opined in 1935. If capitalism was the problem, the Roosevelt Administration's solution was a robust welfare state. "Security" became FDR's rallying cry.

In response, business went on the offense, embracing the concept of security in order to redefine it. As historian Jennifer Klein demonstrates in her 2003 book *For All These Rights*, corporate elites devised a "firm-centered definition" of security "in order to cultivate workers' loyalty to the company and check the further growth of the welfare state" A full-time job became more than just a paycheck—it was a lifeline to unemployment benefits, retirement funds, and medical care. Inadequate public programs left a gap that employers stepped in to fill; security was offered on an occupational as opposed to universal basis.

The advent of neoliberalism in the 1970s and 1980s marked the end of this arrangement. Corporate leaders launched a concerted counterattack, breaking unions, reneging on pensions, freezing wages, dodging taxes, and outsourcing jobs. Determined to erode what remains of the New Deal compact, companies today strategically deploy digital technology to shirk their end of the historic bargain. The development of AI-enabled labor-management systems undermine worker rights and safety through a combination of surveillance and predictive analytics, impacting everything from hiring to firing. Bots scan resumes and assess vocal intonation during job interviews, dictating who gets their foot in the door.

On the job, algorithms are harsh taskmasters, ranking and rating workers and automatically setting performance targets. Pickers in Amazon warehouses and shoppers for Instacart scramble to meet the demands of their digital overlords, enduring mental stress, chronic injuries, and even death. Unpredictable algorithmic wage cuts undermine economic stability, but only come after a worker has been lured to a platform with promises of a living wage and "flexible" hours. Meanwhile, workers are kept in the dark, unable to understand or contest the decisions that set the terms of their lives and livelihoods.

For years, wannabe prophets predicted that robots were "coming for our jobs." That particular forecast, it turns out, has not come to pass. Instead, more and more people effectively have robots for bosses. Aggregating power in the hands of owners, managers, and developers, these digital systems secure profits by disregarding human beings' need for predictable incomes, schedules, and benefits — for security, in other words. When your supervisor is an algorithm, expect no remorse.

Shut Down the Machine

The powerful have never wanted the masses to be secure, and the current crop of Silicon Valley overlords are hardly innovative in this respect. In the Middle Ages, Christian thinkers denounced security as a sin and an insult to God. "Perverse security," they called it. Today, tech billionaires are busy devising new sophisticated tools to spread insecurity so that they might become tomorrow's trillionaires.

Positioned to profit from mass precarity, they know that "disruption" is not a regrettable pitstop on the road to broadly shared prosperity, but a never-ending process that facilitates exploitation. Companies like Uber and Lyft benefit from the fact that millions of people can't make ends meet with one job; from the fact that rising housing costs mean a growing number of drivers have no choice but to live in their cars (many rent their vehicles from Uber, essentially turning them into app-controlled sharecroppers); and from the fact that most drivers are ashamed of their predicament, which makes them less likely to stand up for themselves. It's not a coincidence that homelessness tends to spike wherever the tech industry flourishes. The Bay Area's tent cities are symbolic of a digital economy that thrives on new forms of enclosure and dispossession.

> *"When your supervisor is an algorithm, expect no remorse."*

Looking at capitalism through the lens of insecurity, as opposed to focusing solely on inequality, reminds us that people need more than higher pay; we need peace of mind and an ability to plan ahead. Strong regulations and robust public services are essential in this regard. Newly popular proposals, including national rent control and a homes guarantee that makes affordable housing a universal right, would have a socially transformative effect. In terms of labor rights, AB5, a California law that would classify drivers as employees and not independent contractors, is an encouraging development. Clarifying the employment relationship would help stabilize people's incomes and lives—which is why Uber, Lyft, DoorDash, Postmates, and Instacart have committed $110 million to overturn it.

But even if AB5 became federal law, and "independent contractors" were finally recognized as what they are—workers—that wouldn't be enough. The time has come to decouple security and employment, while also rethinking what security means in an age of ecological crisis and technological possibility. Indeed, the coronavirus outbreak has offered a distressing preview of the sort of upheaval climate change has in store. Overnight, millions of jobs evaporated and countless families were cut off from their medical coverage when they needed it most, laying bare the shortcomings of the New Deal settlement for all to see.

In contrast to the New Deal's individualistic and firm-centered conception of security, we need to devise a truly socialized security system, one predicated on universality and sustainability and geared toward redistributing not just wealth but risk. The frame of a Green New Deal moves us in this direction by centering collective solutions such as public housing, Medicare For All, and a federal job guarantee. (The multi-trillion-dollar stimulus package passed by Congress in the wake of the pandemic proves we have the money to pay for such policies, if only we can muster the political will.)

> *"A new digital arsenal is being developed to ensure we remain insecure despite the abundance in our midst."*

Digital technology can and must be redirected to assist the cause—smart machines could help pair patients with physicians and improve their treatment, for instance, not target desperate customers with discriminatory insurance rates. But digital technology must also recede when appropriate. Leftists these days often say we need to *decommodify*, *democratize*, and *decarbonize* various realms of social life; we need to *de-digitize* many of them as well.

While some technologies—think of the internet service providers we depend on for connectivity and the social media platforms we use to communicate, which have become even more essential to survival as people seek to physically isolate themselves—should be socialized and managed as public utilities, there are other technologies that shouldn't exist, and plenty of data that shouldn't be collected. As tech critic Ben Tarnoff has argued, technology employed primarily for social control or to enforce austerity must be *abolished*, not democratized or socialized. The Atlantic Towers tenants were right—invasive facial recognition systems have no place in our communities—and they proved that popular mobilization can push back on harmful technology. Many more mobilizations will be needed to dismantle insecurity-generating algorithms in all of their forms, from biometric tracking in the workplace to behavioral targeting by advertisers.

Human beings will, of course, never be totally secure. The Stoic philosophers who first pondered the concept understood security as a psychological state, a kind of mental serenity that vulnerable, mortal, meaning-seeking creatures rarely feel. Two thousand years later, we live in a world where, though security on an existential level continues to elude us, economic security, or the fulfillment of everyone's basic needs, is feasible. And yet we inhabit a paradox: a new digital arsenal is being developed to ensure we remain insecure despite the abundance in our midst. Denied the basic resources we need to live, we are forced to seek security through market means—investing in our brands, paying our insurance premiums, and praying that our retirement funds appreciate—while lining the pockets of those most responsible for making us insecure in the first place. Building a more secure world for everyone will be a challenge. But it's a risk we can't afford not to take.

Astra Taylor's latest book is *Democracy May Not Exist, but We'll Miss It When It's Gone*.

Auto Controllers

by Emma Weil

New technologies are giving auto lenders more sophisticated ways to bleed borrowers dry.

T. Candice Smith was driving to work in the express lane of a Las Vegas freeway in 2012 when her car suddenly malfunctioned. Her steering wheel froze and the car shut off. When it rolled to a stop, she and her friend got out and pushed the car to the shoulder. Once she was out of the way of traffic, she noticed a chirping sound coming from inside her dashboard. It was coming from a machine installed by her auto lender: a "starter interrupt device" (SID) that was programmed to make that sound before it disabled her car as punishment for missing a payment. It was then that she realized why she had almost died in a car crash: her lender must have tried to remotely disable her car. So she sat in the immobilized, chirping vehicle on the side of the freeway and called him.

Having her car break down on the freeway was terrifying. Having to wait on the phone with her lender to get her car operable again made her feel even more helpless. When she told him what had happened, the lender insisted that the device wasn't meant to stop a moving vehicle, only to prevent it from starting, and refused to admit any responsibility. Her lender went on to disable her car at least four more times, including once again while she was driving, and then took her to court for not making payments—which later investigations showed the company had actually received but had failed to put in her account. In reality, Smith was never more than a couple of days late on her payments, and never even close to defaulting, which Nevada state law defines as being thirty days past due. And even though it's illegal for creditors in Nevada to repossess a car before the borrower defaults, disabling a car via SID didn't clearly count as a repossession because it doesn't technically "seize" the collateral.

> *"SIDs are not just a potential safety hazard—they are also ruthless tools of financial extraction."*

A year after her near-crash on the freeway, Smith testified about her experience to the Nevada Assembly Committee on Commerce and Labor. Her testimony sparked a wave of national media coverage on SIDs. Calling them "kill switches" and invoking "Big Brother," reporters from the *New York Times* and *Mother Jones* highlighted the cruelty of lenders who would disable a car after just a day without payment. Even the *Daily Mirror* proclaimed, "Heartless creditors LOCKED T. Candice Smith's steering wheel and stopped her engine as she drove down a busy Las Vegas highway."

Smith's testimony was crucial for exposing SIDs to people outside of the auto finance industry, which had already been using them for more than a decade. But her experience was somewhat unique: she was fortunate enough to be able to make the majority of her payments on time, and she experienced an absolutely indefensible near-death collision because of her SID. For these reasons, the media was particularly sympathetic to her.

But there are over two million drivers with SIDs in their cars today, and most of them are not in Smith's situation. They can't always afford to keep up with their payments, and their experience with SIDs aren't nearly as dramatic. Their stories rarely make the headlines, yet these are the people we need to focus on to understand how SIDs really work, and the insidious problem they represent.

SIDs are not just a potential safety hazard—they are also ruthless tools of financial extraction. Lenders prey upon an increasing number of vulnerable borrowers who can't pay back a loan, but who can be squeezed for irregular payments under the coercion of a device that can shut off their car. Lenders see themselves as providing a utility like a phone company, but if anything they're more like the mob. As one dealer remarked, "It's amazing how people manage to pay when they know their car won't start."

There's an App for That

SIDs debuted in the world of auto lending in the late 1990s under the name On Time, offered by a company called Payment Protection Systems. Mel Farr, a running back for the Detroit Lions who became a well-known businessman, helped make On Time devices famous, using them in thousands of his business's auto loans. At its peak, his group of fourteen car dealerships grossed nearly $600 million annually. Farr's business was divisive: he was praised by then-president Bill Clinton for building the nation's largest Black-owned business and "[bringing cars] to every community in this country," but also criticized by others, like consumer protection activist Ralph Nader, for charging incredibly high interest rates—even on cars with the On Time devices.

At the time, the devices followed a simple schematic. There was a small computer, about the size of a billfold wallet, connected to a switch between the ignition and the starter motor, with a voltage monitor on the engine, and small keypad. The computer

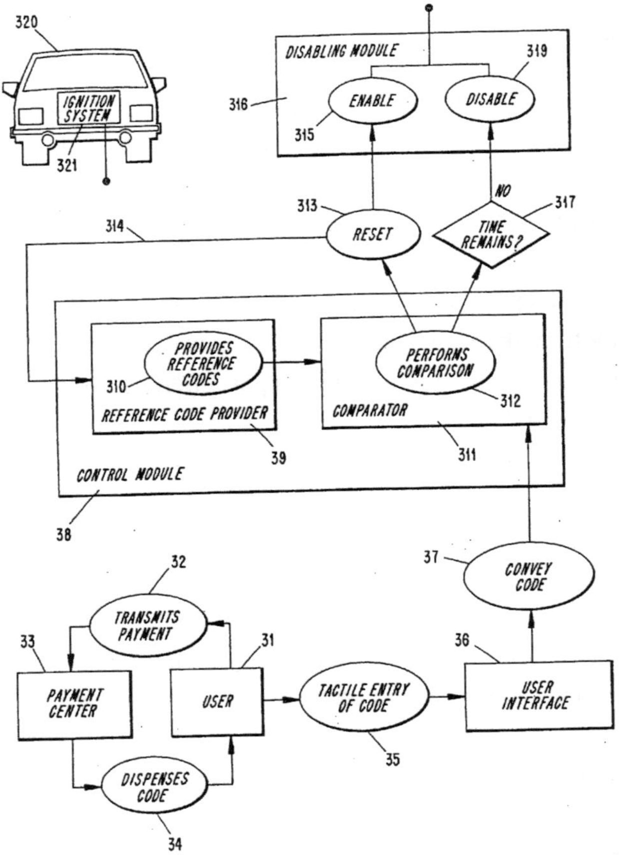

Diagram of a starter interrupt device based loan repayment system, from a patent filed in 1999 by Frank and Mike Simon of Payment Protection Systems, Inc.

was preprogrammed with a schedule and a list of numeric codes, which borrowers would receive from the lender when they made their weekly or bi-weekly payments. If they didn't enter an acceptable code, either a light would flash or a sound would play, signalling that the car would soon be disabled. The voltage monitor on the engine was supposed to ensure that the car wasn't in motion before the computer switched off the starter. In August 1999, On Time's creator, Frank Simon, estimated there were about 15,000 to 20,000 of these devices on the road.

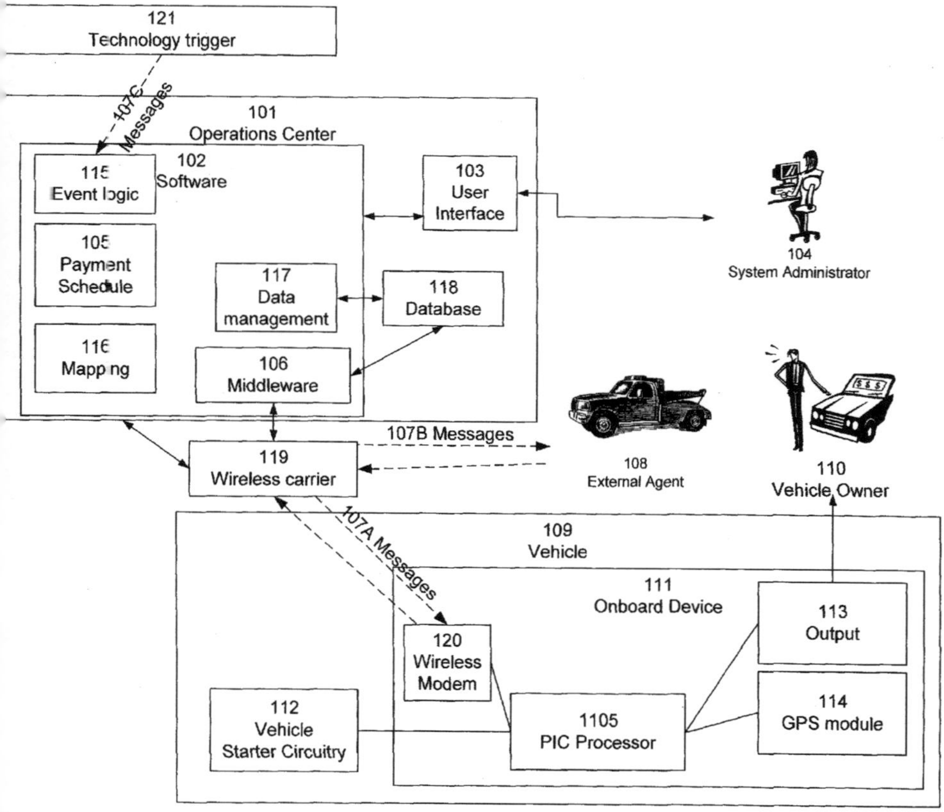

Diagram of a starter interrupt device with GPS and wireless connectivity, from a patent filed by Stanley Schwartz, cofounder of PassTime, in 2006.

The next generation of SIDs arrived in the mid-2000s, which is when the technology really took off. In 2006, Stanley Schwartz, cofounder of a competing SID company PassTime, filed a patent for a combined SID-GPS device. It took advantage of newly robust GPS infrastructure following the launch of the first modernized GPS satellite in 2005. The updated SIDs also utilized a wireless modem, which meant that lenders could communicate with the device in real time through an online portal—and later, on a dedicated smartphone app. Lenders could now disable cars outside of the set schedule, and could even set up geographic boundaries outside of which the car would automatically have its ignition disabled. By 2013, PassTime had sold 1.5 million of these devices—a hundred times as many as the old model.

The new SIDs made disabling a car incredibly easy, both logistically and emotionally. One collections agent told the *New York*

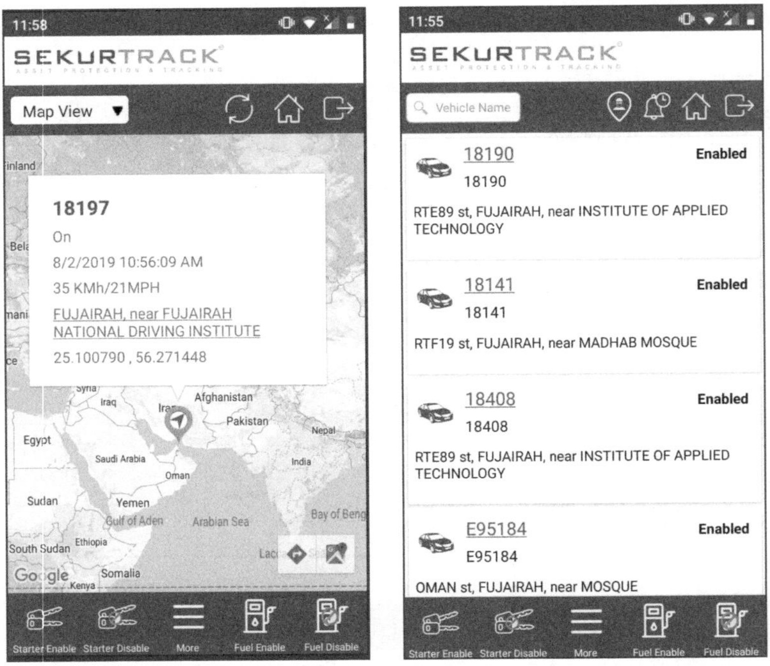

Screenshots of the SekurTrack app on the Google Play store.

Times that he could monitor the location of 880 borrowers' cars from his computer or phone, and that he once casually disabled someone's car from his phone while he was shopping at Walmart.

What about the claim from borrowers like Smith that the SID stopped their car while they were driving? There is no simple answer. But car engines can stall, and if the starter motor is off, then the engine can't restart. Considering the low quality of cars usually offered by no-credit-check used car dealers, it's not hard to believe that the SID could trigger latent engine issues. Or the SID itself could be faulty or improperly installed. Of course, there's a huge legal incentive for lenders to deny these possibilities so they can continue using the devices, because electronically disabling equipment is only legal when it doesn't cause physical harm.

> *"The updated SIDs also utilized a wireless modem, which meant that lenders could communicate with the device in real time through an online portal—and later, on a dedicated smartphone app."*

Target Acquisition

Ever since Mel Farr helped popularize the first generation of SIDs, lenders have stood by the claim that SIDs allow them to fill the crucial role of extending credit to people who would otherwise not qualify for the cars they desperately need to get to work—the so-called "subprime" market. According to *Mother*

Jones, in 2016 about 70 percent of these loans used "payment assurance" devices, which includes SIDs in their early and later forms, as well as GPS trackers without SIDs.

This is a multi-billion-dollar market. Auto dealers target these millions of borrowers with low or non-existent credit scores for loans even without SIDs because, despite the potential for default, they can still make money both by inflating car prices and by charging high interest rates. And the market is growing: In 2019, a record seven million Americans were over three months late on their auto payments, largely due to the increasing number of subprime borrowers.

Theoretically, SIDs could benefit borrowers by helping reduce these interest rates, which should be based on the risk for lenders. But in reality, interest rates are "not [reduced by] installation of the [SID]," according to testimony by Sophia Romero of Legal Aid of Southern Nevada. Instead, SIDs encourage lenders to offer larger loans on higher value vehicles, with the highest possible interest rates. In a 2012 survey of 1,300 dealers, "nearly 70 percent of respondents indicated that they believed the use of devices allowed them to sell higher value... vehicles with less risk." This belief comes from the fact that lenders can continually threaten to remotely disable the vehicle while keeping drivers in their cars, paying for as long as possible until they default—at which point, the lender can quickly repossess the car and sell it to someone else. And in the background, lenders package and sell these predatory loans as asset-backed securities to banks. This should sound familiar—in housing, it's what caused the 2008 financial crisis. And just like in 2008, it's the borrowers, not the banks, who suffer.

So, it's true that SIDs reduce risk so lenders can extend credit to borrowers who otherwise wouldn't get it. But SIDs have nothing to do with someone's actual ability to pay, which means borrowers' pockets are drained while lenders jump indiscriminately from one borrower to the next. Keeanga-Yamahtta Taylor describes a similar dynamic in the housing market as "predatory inclusion," where mortgage lenders in the 1960s targeted previously excluded, mostly Black borrowers, with home loans that were federally insured to reduce lender risk. Instead of actually supporting these borrowers, they raked in as much money as possible until borrowers—who were portrayed as unintelligent and irresponsible—foreclosed.

> *"In 2019, a record seven million Americans were over three months late on their auto payments."*

Privacy-Respecting Predators

So what should we do about SIDs? To date, the legislative responses have been woefully inadequate. They have failed to address the deeper issues of economic justice, and have largely accepted the lenders' basic argument that expanding access to credit is necessary even as it becomes predatory.

For example, in 2017, the FTC started investigating a number of finance companies on their use of SID-GPS tracking devices in the name of consumer privacy. The same year, Nevada passed SB 350, which is one of the strongest state bills regulating SIDs in the country to date. Initially written as a complete ban on SIDs as "deceptive trade practices," the bill was amended to allow SIDs as long as lenders follow certain rules. These rules include requiring written consent from borrowers and certifying proper installation, banning vehicle disablement until thirty days of overdue payment or of operating vehicles, supplying emergency restarts for consumers, and deleting GPS data once it's 180 days old.

> *"We can imagine a world without SIDs, but it's harder to imagine a world without the system they belong to."*

These regulations mainly address security, safety, and privacy—which all must be addressed if SIDs are on the road. But they're not nearly enough to prevent lenders from using SIDs to scale their operations beyond people who can reasonably pay.

One solution would be to force more risk back onto the lenders. For example, states could require lowering interest rates in conjunction with the SIDs. Another would be to restrict their usage overall. States could ban SIDs as a condition for granting a loan—which California actually did, but only for GPS devices, not SIDs. There's also the model of the Consumer Financial Protection Bureau's Payday Lending Rule, which has been indefinitely stalled after its introduction in 2017, but would have required lenders to determine whether a borrower is reasonably able to pay back a loan before granting it.

The path forward is not simple or obvious. Ultimately, the problem with SIDs is that they broaden and intensify a predatory system, so even with laws like SB 350, vulnerable borrowers will still end up trapped in cycles of debt or declaring bankruptcy. As one woman interviewed by CBS News about her experience with SIDs explained her reason for taking out an auto loan at a predatory dealership: "I had to get to dialysis. I have to get there or I don't live." We can imagine a world without SIDs, but it's harder to imagine a world without the system they belong to. Even if we banned the devices, people would still depend on credit for access to life's necessities—and that dependency will always attract lenders trying to make quick money.

Introducing better privacy protections for the data they gather, or ensuring they are safely installed, will do nothing to prevent that. Perhaps we need to "rethink credit as social provision" as the scholar Abbye Atkinson argues, and develop more direct ways of fulfilling people's needs. Perhaps we need to reduce our dependence on cars by providing better public transportation services that make mobility a social right. The fact that people need cars to get to work or the doctor shouldn't be an argument for letting lenders rip them off. It should be a sign that we need a renewed program of investment in our social and physical infrastructure, one that supplies kinder and more creative solutions.

Emma Weil is a research assistant at Upturn in Washington, DC.

Access Denied

by Joyce S. Lee

Online security can be a nightmare for disabled people. A handful of researchers are trying to change that.

Technologists try to make digital experiences frictionless. Mirroring the mythical efficiency of a perpetual motion machine, the goal is to advance users from screen to screen with ease. As a result, standard design practices call for masking complex technical systems beneath simple interactions.

Yet this becomes difficult to do within the context of online security. Security *requires* friction, such as "strong" but difficult-to-remember passwords. Complexity becomes a feature, not a bug. And this complexity tends to increase over time, thanks to the cat-and-mouse nature of security: bad actors develop new tactics that result in high-profile hacks, which in turn demand ever more elaborate countermeasures.

These countermeasures create friction for everyone—but especially for people with disabilities. Indeed, online security can be a nightmare for disabled people. Take the common security practice of converting letters to asterisks (*) in password fields. The screen readers used by visually impaired people try to mirror this practice by replacing every occluded password letter with the word "star" or calling the entire field "concealed text." Imagine for a moment typing in your password, and hearing either "star" repeatedly or complete silence. This makes it difficult to review what letters have been typed, leading to mistakes and frustration.

Another example is facial recognition for devices, like Apple Face ID and Windows Hello. In addition to performing inconsistently for underrepresented groups, facial recognition creates new challenges for blind users. "Ninety-five percent of the time the camera can't see me," says Lucy Greco, the web accessibility evangelist at UC Berkeley. "I know where to touch the fingerprint sensor because it's the same spot every time. But I don't know why it's not seeing my face."

These kinds of difficulties cause many disabled users to opt out of security measures entirely. "I have noticed that many people who are emerging technology users—who are blind, visually impaired, or have other disabilities—they'll get a lot of assistance to set up their phone," says Chancey Fleet, assistive technology coordinator at the New York Public Library. "Often they decide, or have the decision made for them, to skip the passcode," she continues. "Because when you're new and everything's hard, entering the passcode is hard. And that lack of passcode will linger for months, or even years, and become habitual."

The result is a crisis. We are creating a digital world where the most desirable path for disabled people is to leave themselves unprotected. This means that as the level of security rises among the nondisabled, disabled people are becoming more vulnerable to digital threats, deepening disparities. And these disparities affect an immense share of the world's population—over a billion people have some form of disability, according to the World Health Organization. Furthermore, this number is constantly growing: disabilities are correlated with age, and lifespans are increasing. We will all be disabled in one form or another during the course of our lives. It's just a matter of time. So when those of us who are (currently) nondisabled contemplate accessibility, we aren't just serving our fellow humans with disabilities in the present—we're also serving our disabled selves in the future.

Fortunately, there's nothing inevitable about a digital world that fails to protect disabled people. Academic researchers are proving that it's possible to rethink authentication for disability, synthesizing perspectives from the realms of both security and accessibility. Their efforts offer a glimpse of what our technology could look like if it aspired to keep everyone safe.

> *"We are creating a digital world where the most desirable path for disabled people is to leave themselves unprotected."*

Open Sesame

To help disabled people remain secure online, academics are exploring solutions involving gestures, haptics, and even "mind reading." Consider the "passthought," proposed in 2005 by computer scientists at Carleton University. Passthoughts involve thinking a silent and inherently unobservable thought. In lab studies, researchers provide instructions to subjects such as, "Imagine that you are singing a song or reciting a passage for ten seconds without making any noise." Detected by sensors, the brainwave signals emitted from this thought can log users into services or devices.

How does this work? Prior to authentication, users are asked to think the passthought repeatedly. Their digitized brainwave signals—the "training data"—are then processed to extract a set of features that best encode a user profile and remain consistent over time. Finally, this set of features is used to build a "lock" that only opens for the specified user performing the specified passthought.

Passthoughts hold promise because they provide two factors of authentication in a single step. Knowledge, or the secret thought, is one factor; the user's brain is the other. Given that each of our brains is uniquely differentiated by genetics and environmental conditions, our brain waves have a kind of biometric signature. But unlike fingerprints or faces, passthoughts can be changed if they are compromised. And even if the attacker knows the content of the passthought, it can't open the lock without the user's brain. This makes passthoughts relatively robust against impersonation and better protected from social engineering attacks, where the user is tricked into revealing their password.

Despite these advantages, passthoughts still have practical limitations for real-world use. In an exploratory study with blind people, researchers found that authenticating via brain-computer interface was less accurate and took longer than entering a PIN. Moreover, the sensor-embedded headsets that are required for brainwave authentication can be awkward, uncomfortable, and stigmatizing to wear. Researchers have explored discrete, in-ear alternatives, but many involve bands that wrap around the neck and place electrodes along the scalp, while more robust models resemble a swim cap. These headsets can also cost hundreds of dollars and so may be out of financial reach for many people.

Another creative approach is PassChords, developed in 2012 by Cornell professor Shiri Azenkot and her colleagues. "These researchers recognized that if you're using a screen reader in public on your smartphone, people might be able to hear you enter your password," explains Cynthia Bennett, a researcher at the University of Washington. "So they came up with a gesture-based system that was silent." Entering a PassChord involves tapping several times on a touch surface with one or more fingers, reminiscent of playing a musical sequence. A study with sixteen blind participants found that using PassChords was as accurate and significantly faster than entering an iPhone passcode with a screen reader, taking about a third of the time.

While passthoughts and PassChords have the potential to improve the lives of millions of disabled users, it will be hard to make them mainstream. Academic researchers are constrained by limited resources. They can build novel prototypes, but they often don't have the capacity to bring those prototypes to a mass audience. "PassChords is heavily cited, it's well known in the community, and in user testing, people say great things about it," says Yang Wang, a professor at the University of Illinois at Urbana-Champaign. "I don't think it's widely deployed though."

Without an expensive marketing campaign, how can a product find users? And without a business model, how can operating costs be sustained? Maintainability is another challenge: projects like PassChords need to be continually updated as browsers and operating systems change. This is hard to do, especially as the students who write the software for these academic projects graduate. Furthermore, in the absence of a revenue stream, projects can't hire people to maintain the software on an ongoing basis.

Academia's distance from industry is thus a double-edged sword: while academics have the freedom to prototype projects that are insulated from market forces, this also limits the scale of their impact. Solutions like passthoughts and PassChords are only available to a select group of users, often for a limited amount of time. They remain in beta indefinitely and are rarely released to the wider public.

Hack the Planet

Unfettered by the resource constraints of academia, tech companies could pick up these projects and deploy them at scale. So why don't they? Because the industry largely considers people with disabilities to be anomalies—a niche and "non-normal" subgroup of users. By this logic, it seems unprofitable to develop specialized features that appear to benefit only a select few. When businesses do prioritize accessibility, they often do so for two reasons: to avoid discrimination lawsuits and to generate positive publicity. The benefits of either are difficult to measure, however, while the costs of building features for disabled users are concrete and non-trivial.

But there is another urgent reason that industry should care about accessibility: security. When companies create bad accessibility features, it can result in more than just bad user experiences. It can also introduce new security risks, impacting systems and individuals alike. And these risks can lead to breaches that inflict real damage to the company's bottom line.

A good example is CAPTCHA, or the Completely Automated Public Turing test to tell Computers and Humans Apart. Designed as a challenge to prevent bots from spamming a system, CAPTCHAs were first introduced as images of distorted character strings. Typing the contents of this image into a text field would effectively prove one's humanity and allow a user to proceed onto the next webpage.

> *"The industry largely considers people with disabilities to be anomalies — a niche and 'non-normal' subgroup of users."*

However, the original CAPTCHA was soon revealed to exclude people with visual impairments, effectively deeming only those who could see images as "human." Including descriptive "alt" text in the source code of the website could solve this problem by making CAPTCHAs accessible to screen readers — but at the cost of also making CAPTCHAs legible to bots, which could then pass for human. Enter the audio CAPTCHA, which was introduced as an "accessible" alternative — though it still excludes people that are deafblind. It reads the contents of a visual CAPTCHA aloud without relying on alt text, and it adds background noise or distortion that makes it harder for bots to understand.

While visual CAPTCHAs have evolved to more robust cognitive tasks — like selecting all the traffic lights in a given image — accessible CAPTCHAs haven't progressed nearly as far. Meanwhile, advancements in speech recognition technology are approaching human performance, giving hackers enhanced capabilities to decipher audio CAPTCHAs even with significant noise or distortion. After obtaining and preprocessing the audio, attackers can now automate the submission of these recordings to speech recognition services, and then plug the transcriptions into the web form. The result is new system-level security risks: in a 2017 study, researchers at the University of Illinois at Chicago found that off-the-shelf speech recognition services could be used to hack the most common audio CAPTCHAs, breaking Google's reCAPTCHA with a 98 percent success rate.

Accessibility done badly can make the internet less safe for everyone. If companies can't be convinced to protect disabled users because of ethical and legal reasons, perhaps they can be motivated by the security risks that result from serving disabled users poorly. Moreover, the potential impact of these risks is growing as accessibility permissions are used in unexpected ways: to deliver new features, rather than expanding access. Even the nondisabled may therefore be unwittingly using accessibility features—and bearing the associated risks.

For instance, Dropbox users are asked to turn on accessibility during setup on Mac computers. Yet how this helps people with disabilities is currently not discussed in the documentation; the explanation emphasizes that turning on accessibility will enable added functionality. "Mac shoehorns a million different things into accessibility permissions," says an anonymous commenter on r/MacOS. "You might grant a window manager accessibility permission to move windows around, but then the window manager could, if it wanted to, read your keystrokes."

Moving Targets

So what is the way forward? Ideally, we could foster partnerships between academia and industry that combine the interdisciplinary prototyping of the former with the funding of the latter. Even under the best arrangement, however, a deeper difficulty will remain: both security and disability are moving targets. What one needs to be secure online is constantly changing—and so is what defines one as disabled. As a report from the AI Now Institute notes, "the boundaries of disability... have continuously shifted in relation to unstable and culturally specific notions of 'ability.'"

> *"Accessibility done badly can make the internet less safe for everyone."*

Security and disability are fluid, and this will always create challenges for technologists trying to design for either—let alone both. Much creative experimentation will be needed, but the goal should always be the same: to keep as many people as possible safe, even if it means overhauling our technologies. As Bennett, one of the authors of the AI Now report, explains, "Disabled folks are not vulnerable because of our inherent bodyminds, but because systems are not set up to protect all people equally. That's what creates the vulnerability."

Joyce S. Lee is a UX researcher based in Oakland, California.

Safe or Just Surveilled?

Tawana Petty on the Fight Against Facial Recognition Surveillance

Organizers across the US have been building a movement to ban police use of facial recognition technology. To date, they have won municipal-level bans in San Francisco, Oakland, and Berkeley, as well as in Somerville, MA; however, the struggle against surveillance technology continues elsewhere.

We talked to Tawana "Honeycomb" Petty about what that fight looks like in Detroit, a majority-Black city where global capitalism and municipal disinvestment have left many Black residents below the poverty line. For decades she's worked in the lineage of James and Grace Lee Boggs and other community anti-racist organizers to remind Detroit residents that Black abundance and safety existed before and are possible again.

Today, Petty organizes against surveillance technologies in Detroit as the director of the Data Justice Program for Detroit Community Technology Project. She also co-leads Our Data Bodies, a five-person team that raises consciousness in marginalized communities about how digital information is collected, stored, and shared by governments and corporations. She is a convening member of the Detroit Digital Justice

Coalition, which organizes "Data DiscoTechs" and other initiatives to foster media and digital literacy.

The work is all interconnected. How do you equip residents with what they need to demand what's best for them and their communities? Whether helping people to understand surveillance, advocating for inclusion in the first digital Census, or navigating government benefits, Tawana's work with community members focuses on education, organizing, and agency in the pursuit of digital justice in Detroit and beyond.

We sat down with Tawana in January 2020 to talk about the fight against facial recognition, building data agency within systems meant to strip it, and what's at stake in what she calls "the last Black mecca" in the United States.

To start us off, could you tell us about yourself, your background, and how you came to work with the Detroit Community Technology Project?

I'm Tawana Petty, also known as Honeycomb. I'm a poet, a mom, a social justice organizer, and a water rights advocate. I direct the Data Justice program for Detroit Community Technology Project (DCTP), and through DCTP I convene the Detroit Digital Justice Coalition.

Right now, one of the Data Justice Program's main initiatives is slowing down the rapid expansion of Project Green Light, which is a public-private partnership around facial recognition technology spearheaded by the Detroit Police Department (DPD). We're also creating educational resources about technology, surveillance, and equitable participation in the Census. During the time that Project Green Light was ramping up, DCTP was involved in a research project called Our Data Bodies, which I can tell you more about later.

When Project Green Light got started, we had no concept of the scope of surveillance it would take on. The project gets its name from flashing green lights that are connected to video surveillance cameras inside and outside of different businesses. The cameras are monitored twenty-four hours per day, seven days per week, at DPD's real-time crime centers and mobile devices.

It started off with cameras at eight gas stations—ones that stayed open during late night hours. The police wanted to use these cameras to signal to community members that the gas stations were now safe to enter at any time, because police were constantly going to be watching them. They partnered with two private companies on this project: Guardian Alarm and Comcast.

Fast forward a couple of years. We now have close to six hundred cameras all over Detroit, and the Mayor would like to push that number to 4,000. Project Green Light locations pay a monthly rate so that if something happens at that location, they get priority from police over non–Green Light locations. So they pay for policing. And then, of course, the DPD leadership signed a contract to use facial recognition on everything from drones, traffic lights, mobile devices—pretty much anything they could attach a surveillance camera to. They were using that facial recognition technology on footage from their Green Lights for about two years before social justice activists and technologists in the city really got wind of it, when DPD tried to push through a directive that would solidify their use of it. Before that proposal, there was no policy governing its use.

What did the rollout of Project Green Light look like within the community, and how did you start to organize folks against it?

The police really tapped into this negative narrative that has hovered over Detroit for decades. They were touting Project

Green Light as the salvation of the city, like, "Hey, we have the answer. We know you are all afraid of each other, so we're going to give you all these security cameras, and you're going to feel safe because we're always going to be watching you."

The police department's campaign targeted senior citizens mostly—the ones who are retired or sitting at home, inundated with media images. They didn't hide the idea of putting cameras everywhere like they later hid their adoption of facial recognition. In fact, they were trying to separate the two ideas, saying that Project Green Light was different than facial recognition. But the two systems needed each other; the city needed to overhaul itself with all these surveillance cameras in order to make facial recognition a viable system. The DPD just kept saying, "Facial recognition is not embedded within the cameras," and, "We'll only use the facial recognition if we absolutely have to." Meanwhile, they were proposing policy directives that were asking for real-time live tracking on mobile devices, drones, traffic lights, and more.

Initially, when we were showing up to meetings at the Board of Police Commissioners, there were a lot of residents, mostly senior citizens, who were really angry with us. They felt that we were getting in the way of a system that would make them safe, and that we were attacking police because we just wanted to cause trouble. In reality, we were just trying to educate residents, law enforcement, and the Board of Police Commissioners on the harms of mass surveillance and facial recognition. Over time, we were able to argue that facial recognition and Project Green Light needed each other to work the way that DPD leadership and city government wanted them to work, and you couldn't have one without the other. We were also able to make the case that safety and surveillance were not synonymous.

Was your resistance mainly organized around police commission meetings? How else did you all create pressure?

We spoke at police commission meetings and town halls. We collaborated on a special surveillance issue of a magazine called *Riverwise*, where we brought together information about Project Green Light, the problems with facial recognition, and alternatives that would create actual safety. We also created a report, "A Critical Summary of Detroit's Project Green Light and its Greater Context."

"Green Chairs, Not Green Lights" is our counter-campaign. We're encouraging community members to come back to their front porches and look out for each other. We're raising consciousness and building real safety, and we're saying that a system set up to prioritize profits and institutions over people is wrong.

I imagine you all are working in coalition with many different groups in this campaign. What are those different groups? Also, there's a long history of activism and in particular Black-led activism in Detroit. How do you and the rest of the coalition situate yourselves within that history?

Yeah, absolutely. In addition to serving as director of the Data Justice Program here at DCTP, I'm also on the board of the James and Grace Lee Boggs Center to Nurture Community Leadership. James and Grace Lee Boggs are now ancestors, and a long legacy countering these sorts of things was created under their expert tutelage. [*Eds.: James and Grace Lee Boggs were revolutionaries who worked to dismantle capitalism and racist oppression nationally and through fostering community-based, grassroots activism within their adopted home city of Detroit.*] Our former Boggs Center board member Ron Scott, who is also now an ancestor, co-founded the Detroit Coalition Against Police Brutality, which still exists. I'm part of the Black Out Green Light Coalition and

Green Light Black Futures coalition, which is organized by Black Youth Project 100 (BYP100). I also collaborate with the Detroit Justice Center, Media Justice, Color of Change, and Fight for the Future. Finally, I'm part of a lot of coalition work with the ACLU that collaborates with dozens of organizations that have been consistently resisting mass surveillance in the city.

Ultimately, we were unsuccessful in getting DPD to stop using facial recognition, but we did have some wins within the policy that the Board of Police Commissioners ultimately passed. We're definitely tuned in, which is why I think we were able to achieve any policy changes at all, even if what we wanted was a ban. Honestly, getting any of those changes in a city with this many Black people took a tremendous amount of work and was an uphill battle. All the other cities that were successful with bans were predominately white. Folks aren't as eager to take the ever-watching eye off of a predominately Black city.

Could you tell me about those changes and what the policy looks like now?

Under the new policy, the police are prohibited from using facial recognition on any live feeds. Even though still photos are also very problematic, the thought that DPD would have been able to track people's faces as they walk the streets was very unnerving. So they're not able to do that. DPD can only use facial recognition to investigate violent crimes. They're barred from using it for immigration enforcement. They're barred from predictive analysis—using the technology for pre-crime, essentially. They're barred from mobile use, so they're not able to use their mobile devices to walk through crowds and track people using facial recognition, which is something that happened in Baltimore during the Freddie Gray protests. They're barred from violating the First, Fourth, and Fourteenth

Amendments—freedom of religion, prohibition against unreasonable search and seizure, and due process under the law. Now, I'll tell you that I don't believe that's even possible—I think if you mass surveill a city, there's no way you're not going to violate those amendments.

The Board of Police Commissioners receives a weekly report on how the police are using facial recognition. Unfortunately, the Board of Police Commissioners, which is supposed to be a civilian oversight body, hasn't been very transparent with those reports. None of them have been made accessible to the public. They don't even question the contents of the reports when they are handed them in meetings, at least not any of the Board of Police Commissioners meetings I have been present at. It's discouraging.

Most significantly in the policy, Detroit police officers are supposed to be criminally investigated if they misuse the technology. Their misuse is supposed to be reported to the mayor, the Board of Police Commissioners, and the city council within twenty-four hours of discovery, which would lead to dismissal and further legal action.

But I would say that our greatest accomplishment was raising the consciousness of the city residents. Many, many residents were engaged in a process that, at first, they were just going along with.

How did you do that?

For several months, we gave testimony at every Board of Police Commissioner meeting that involved the topic. Unfortunately, residents are only afforded two minutes to speak, even on topics as dire as this one. We leveraged our minutes to push forward a vision of people sitting on their front porches, the community having good schools, affordable water that doesn't contain lead, neighbors not getting evicted, neighborhood block clubs. We pushed for senior citizens to think about the times when they felt safe, and whether or not being held up in their homes under surveillance cameras actually made them safe.

> *"It's going to be much harder to convince not just Detroiters but the world that a Black city doesn't need to be surveilled."*

That understanding of safety was the scaffolding. And once we had that, we could build on it. We could say: what the police department is talking about is *security*; it's not what you're thinking of when you're thinking about *safety*.

It really was a slow educational process, to be honest with you—constantly bringing materials and giving testimony. We'd bring examples of conversations we'd had with children, where we'd ask them a question like, "Say you lived on a street. On side A of that street, there were metal gates and cameras and drones and police. And on side B, there were a bunch of neighbors who knew each other, who looked out for each other, who knew that if John didn't come home at 7:00 p.m., then they should probably find out where John is. On which side of the street would you feel safest?" Kids almost always said, "How could I feel safe if I don't know the people who live next to me, you know?" That same type of teaching works with adults when they started to think it through.

We also began bringing in the data around how facial recognition technology misidentifies darker skin tones in women and children, and how Detroit is 80 percent Black. We started reflecting back what DPD leadership themselves were actually saying. For example, DPD leadership indicated that of the 500 times over two years that they used facial recognition, they only moved forward with 150 cases, because the other 350 were misidentifications by the system.

We highlighted the fact that they were telling us the system was wrong at least 350 times and that they were relying on the naked eye of two analysts to catch the errors of the algorithm. This caused great concern about the potential of false arrests. So, we dissect what's being said and present that back to people consistently.

You've mentioned how Detroit is 80 percent Black, a stark contrast to other, primarily white cities that have banned facial recognition. How does your fight in Detroit look different from the fight in whiter cities like San Francisco or Boston?

We've built relationships with organizers in pretty much all the other places that have successfully banned facial recognition, including Portland, where folks are pushing for the most comprehensive ban. We've consistently told them: we need to synthesize what you all have done, but also apply a racial lens to it because it's going to be much harder to convince not just Detroiters but the world that a Black city doesn't need to be surveilled. It's a deeper, larger conversation and it requires a kind of organizing around anti-Black racism that often isn't comfortable to do even in social justice or liberal spaces.

That's why I value the consciousness-raising piece so much. I really want to win this, but I also have to find some value in the fact that we've been able to raise the consciousness of people all across the US who now consider themselves invested in our struggle. So I take pieces from all the places that I've been able to visit and all the people that we've been able to talk to. But I also understand that there is a very real dynamic here that does not exist in cities that only have a 5 percent Black population.

One of the things that I've consistently said is that Detroit is the last remaining Black mecca in the United States. If we don't succeed in resisting some of this violent system, there is no hope for Black residents across the rest of the country, period. Police departments have already tried to model this system in other cities—calling them Project Blue Light in Wisconsin, for example. They're just packaging up what city government is pushing through in Detroit. On any given day here, you can see drones flying through the sky like kites.

Wow, that's an apocalyptic vision. In past talks, you've talked about the rise of surveillance and how that ties in with the "resurgence" of the city, as money floods into gentrified areas like downtown and Corktown. Could you give a little bit of that historical context in Detroit?

There are areas of Detroit like Black Bottom and Paradise Valley that had Black hospitals, Black grocery stores and schools, and African-centered education—all these viable institutions—that were leveled for freeways. One of our challenges is getting folks to think about a time in the city when there was an abundance of thriving neighborhoods that were predominantly Black-owned, -led, and -invested.

I'm always trying to point out the inaccuracy of the narrative that Detroiters just don't want to care for where they live, they don't want safe neighborhoods, they don't want viable institutions—that the system that we exist under is a *choice*.

We're in a city with five hundred thousand Black people and the median household income here is $29,000 a year—which means that half the households in the city make less than that. Predominantly Black women-led households. Lots of children in extreme poverty. One hundred thousand people without water in their homes over the last ten years. Schools closed by the hundreds. I work to get people to see crystal clear that this isn't an accidental situation. It's easy to convince anybody in the state of Michigan and the world that if something bad is happening to Detroit, we brought it on ourselves. There isn't a dissecting of how we got here, or the history of racism and the disinvestment that has happened for generations, you know?

> *"Detroit is the last remaining Black mecca in the United States. If we don't succeed in resisting some of this violent system, there is no hope for Black residents across the rest of the country."*

Watched, Not Seen

At the beginning of our conversation, you mentioned the Our Data Bodies project. I understand the project is led by a five-person team that includes three organizers from different groups: the Center for Community Transitions in Charlotte, which focuses on folks who have recently reentered the community from incarceration; the Stop LAPD Spying Coalition, which organizes against police surveillance

in Los Angeles; and you all at DCTP. How did those three organizations come together?

Our Data Bodies was initially a two-year research project between Detroit, Charlotte, and Los Angeles. We were thinking about the digital streams we all produce and the impacts that our information has on our interactions with government and police, on how we get resources, and on how our cities are developed. Within those two years, we learned that we wanted to move beyond a research project and expand on our organizing within our communities to increase self-determination around these systems.

The three organizers—Mariella Saba, Tamika Lewis, and I—all originally applied for our positions on the project. The interview committee was very intentional in picking social justice organizers for these roles as researchers because the type of analysis that comes from organizing on the ground in the neighborhood was going to be important in making sure that we were prioritizing community members. That won't happen if you just focus on having a good researcher. Research is something that can be taught; care and compassion for the people connected to the research is a longer process.

So we all play to our strengths. I consider Stop LAPD Spying a mentor in this work, especially in resisting police and surveillance systems. For this project, Stop LAPD Spying and Los Angeles Community Action Network (LA CAN) put a lot of emphasis on unhoused populations and how Skid Row is a heavily targeted community that the LAPD uses to innovate new surveillance systems.

In Charlotte, Tamika does work around citizens returning from incarceration. Folks returning from incarceration constantly have to report back to the system. It's like the system is constantly waiting for you to fail. Tamika also brings an important perspective as a gender-nonconforming organizer: they understand how resistance to being defined under the dominate gender binary makes people targets for intensified surveillance.

How have you been able to make use of that experience and research?

We've collectively produced materials like our Digital Defense Playbook, which is an educational resource and activity guide about data, surveillance, and safety that we created last year based on our three years of community research. We've taken that work to different neighborhood institutions, as well as academia. We've worked with data scientists, who are often disconnected from the real people represented by the statistics they analyze.

We're all learning from each other and staying in constant communication about what's happening in our different cities, and hearing the same themes. Things like, "The one mistake that I've made in my life is now the thing that's limiting how I'm able to survive." Or, "I'm feeling heavily surveilled because I'm trans," or "I'm feeling super targeted because my credit isn't good," or "I've just reentered society from incarceration and I feel like every move that I make is being scrutinized and monitored." Across all the cities, community members wanted to be seen for something more than a mistake they made, or something more than their data. And they didn't want to be pursued and tracked and monitored.

What exactly is a "data body"? Recently, we've seen the rise of various tools that let you see the data that companies have about you. But your approach seems more holistic. It's not just, does Facebook have this data about me, but what are all of the different pieces of data that the government or private companies might have—and what am I able to do about it?

We try to get community members to think not just about the impact that their data has on them, but the impact that their data has on the decisions that affect their family, their neighborhood, and their city.

One of our exercises is called "What's in Your Wallet?" As part of that exercise, we often use the example of a person who uses their EBT card at the liquor store up the street and purchases foods that aren't considered healthy. Those activities create a data trail. Maybe a bank will look at those data trails and decide not to invest in a grocery store in that neighborhood. It's difficult to know exactly how some decisions are made, but what we do know is that data is leveraged to make most of those decisions.

That's what we mean by data bodies. It's not just the individual; it's the information that's been generated about this individual and the systems that interact to make decisions about this individual, this individual's family, this individual's neighborhood—all the data's tentacles.

> *"Power is going to come from the respiriting of community members."*

In the Digital Defense Playbook, you cite a conversation with a community member who says about interacting with government agencies, "For their benefit they do communicate. But for my benefit, no." And I see this a lot in government work—if agencies want to communicate to surveil and penalize, then they can and will. But if you want them to share information to, say, verify that you've lost your job so that you can get food stamps, they often can't or won't.

It seems like so much of your work with Our Data Bodies is trying to build data agency within a system where power is skewed heavily towards other organizations or towards the state. How do you help people build data agency to overcome that power imbalance?

One of the things that we truly, clearly recognize within our work is that the power is going to come from the respiriting of community members. A lot of folks have been really dehumanized. There's a quote that says, "Friendship begins at the moment when one person says to another, 'What, you too? I thought I was the only one.'" And so it's been this weaving together of stories to let community members know that they're not alone.

It's saying: did you know that there is an open data portal in your city and that you can push for that to become a useful source of information? Or: did you know that you can request to have some of your information deleted from a government database? It's letting people know: here's how these systems connect and interact with one another. It's an ongoing process of learning how different people are experiencing these systems, but also tying them together with the stories of others so that they understand they aren't isolated, that there are many people across the world resisting these systems.

We intentionally did not publish everything in the Digital Defense Playbook. There are stories that community members shared with us that emphasize how they're resisting systems that we did not share, which contradicts everything that a capitalist system would ask you to do. It's like, why not put it all out there? Well, because we actually want people to keep surviving.

The Digital Defense Playbook serves as a community organizing mechanism where we get to talk to people, we get to share stories, we get to respirit, we get to raise digital and media literacy. We don't think that it solves all the things. But we do think that when you build the confidence of community members, and you let them know that they're not alone, and that there are folks that are resisting these things all over, it does something for the spirit. So that's how we use it.

In Person, In Print

Like Our Data Bodies, the Data DiscoTechs organized by the Detroit Digital Justice Coalition seem like they aim to raise consciousness and agency around data within communities. Could you paint a picture of one for me?

The "discovering technology" fairs, which we refer to as DiscoTechs, are kind of like a science fair at school—you know, with informational stations on different topics. To start one, we either respond to folks who have reached out to us or pick a neighborhood where we want to do one. We connect with an organizer or resident who lives in that neighborhood, who has the pulse of what's happening, and we support them in finding out what stations they think might most benefit their neighborhood.

We start with a few stations that we definitely want to have. Recently, we've had stations where we're telling community members what to expect with the census. What kinds of things get funded based on census data? What are some of the concerns that have been raised about the census? How has it been used in the past? How long is census data kept? We did research to empower community members with that information.

But then the rest of the eight or nine stations are designed around what the community says they need. So they might say, well, in this neighborhood, it's a lot of senior citizens. They don't understand Twitter, they don't understand how to set up an email account, they just want to learn how to use a Mac—things like that. So then those are the stations that we will have. Community members are able to come in for free. We have music. We call it a DiscoTech, but it also has music and dance. At the last two, we gave away low-cost computers: the Raspberry Pi full kit, with the keyboard and everything. We gave those to residents who don't have that technology at home.

Who shows up to those DiscoTechs? Mostly young folks? Is it a mix of ages, of races?

It definitely depends where we have them. Some neighborhoods are predominantly senior citizens, so we'll have a lot of seniors show up. When we did one in southwest Detroit, it was a predominately Spanish-speaking undocumented community, so we had interpreters onsite, including youth interpreters. It was great. Those stations had mostly young and middle-age participants.

I'm curious about the print magazine that you help produce, *Riverwise*. How are you involved? How did it come to be?

I think it's the poet in me, but I don't remember a point in my life where I didn't feel obligated to push back against the dominant narrative of Detroit. I'm pretty sure there was a time, but I don't remember it.

There was a newspaper in the city, the *Michigan Citizen*, that was trying to lift up community stories of resilience and resistance, countering the dominant narrative in the city. It was around for three decades, but then it ended. The Boggs Center felt that we had a responsibility to fill that void, or at least try to fill that void. And so we convened community members over a one-year period to see whether there was interest in trying to create consistent literature that would amplify community stories of resilience and resistance that would counter the narratives that were coming out of the city. Through those meetings, *Riverwise* was born. I am now part of the Riverwise Magazine Collective, and I serve on the editorial board. Eric Campbell, a former contributor to the *Michigan Citizen* is the managing editor. We put it out quarterly, and then we do some special editions, depending on what the situation is. That's how the special surveillance edition in collaboration with DCTP came about.

We convene weekly and go through stories that we've heard ourselves, or we've pursued folks who we want to write for us, or we have people who send in submissions. Then we determine whether we want to theme the magazine based on something that's happening in the city, or if this is one where we're just going to lift up various stories. And then we have these community discussions around what's written in the magazine. If a person did an article or something for the magazine, then we'll ask them if they want to host a community discussion and we will support them in that. We also host writing workshops to support residents who want to improve their writing skills or learn how to contribute content to the magazine. It's not our coalition that does most of the writing workshops. We tap people in the community—poets, writers, you know, organizers—who want to host those workshops and then we just support them.

Calling All Co-liberators

How can people support the fight against facial recognition?

They can support organizations like Fight for the Future which are pushing for a federal ban on the government use of facial recognition. They could call or email their Congressperson. They can help us amplify our struggles.

In Michigan there are a few current bills that could use some support. There's a bill in the State Senate, 2019 SB 0342, that's likely going to be passed. It was initially a ban on all police use of facial recognition, but now it's a ban on police use of *real-time* facial recognition. There was House Bill 4810, a five-year moratorium on the police use of facial recognition on everything—drones, body cameras, any surveillance technology—that's still kind of lingering out there. And then there is the Community Input Over Government Surveillance ordinance that's before the Detroit City Council, which creates guidelines for council oversight and public input for surveillance technology. That one is currently being passed back and forth for amendments.

On the federal level, there's a Congressional bill called the No Biometric Barriers Housing Act of 2019 that bans the use of biometrics in public housing. There's also a bill out there called the Ethical Use of Facial Recognition Act that would prevent federal funding from being used for facial recognition technologies, including police and other government use. I'm not really sure what's going to stick beyond the Detroit city ordinance, though, and maybe the Michigan State Senate bill.

But wherever you are, you can support the fight by continuing to try to politicize your community, push for alternatives to surveillance, and minimize the conflation between surveillance and safety. And in Detroit, we'll just continue the work by showing how invasive and insidious this technology is—hopefully before we see a slew of false arrests. We don't want to be living in a social credit system, where our every move is dictated by algorithms and surveillance cameras.

Or at least more than what we have already.

Exactly, yeah. It's already a social credit system in Detroit, but we don't want to add more and more technologies to that and exacerbate the violence and marginalization that residents are already feeling. And we don't want this rolled out all across the globe as a way to contain and control Black and brown people and Indigenous people and poor white people. Surveillance is not safety.

"We need co-liberators; we don't need any more allies. We need folks to really feel their liberation is tied up in our liberation."

The hardest part for us is that law enforcement, government institutions, and too many people don't seem to see an alternative, especially in areas that have been deemed dangerous. But everyone needs to understand that most crime in our neighborhoods is rooted in poverty and disinvestment and racial violence. We need co-liberators; we don't need any more allies. We need folks to really feel their liberation is tied up in our liberation, you know?

Tracing Paper

by Mitch Anzuoni

Color printers mark printouts with barely visible codes that are used to track down currency counterfeiters, as well as everyone else.

In 2017, when a National Security Agency (NSA) whistleblower wanted to extract classified government documents from her work computer, she sought refuge in the printed page. Maybe she thought physical paper would be safer from digital surveillance than an email. So she printed the documents at her office and then mailed them to *The Intercept*, which broke the news with the headline, "Top-Secret NSA Report Details Russian Hacking Effort Days Before 2016 Election" on June 5th, 2017 at 3:44 p.m. eastern time. A few hours later, the US Department of Justice officially announced their arrest of Reality Winner, a former US Air Force officer and NSA contractor.

What happened? *The Intercept* contacted the NSA on May 30th asking them to verify the documents. But by sending the scanned images that included each page's wrinkles and folds, as opposed to retyping the information, the journalists shared more than they intended to: they sent the NSA the pale yellow tracking dots that are embedded in every piece of paper that is printed by a color laser printer. The dots form rectangular grids of rows and columns, with each dot's position corresponding to the value of a date, time, or printer model. Together, the rows and columns constitute a machine-readable bitmap known as a machine identification code (MIC). MIC grids repeat across the page so that even if only a shred of a page is recovered, the MIC on that shred can still be decoded and traced. While neither the Justice Department's nor the FBI's statements about Winner's arrest mentioned MICs, security experts strongly suggested that they played a role in helping the agencies identify her and, at the very least, corroborated other evidence linking Winner to the leak.

Winner's case spurred a renewed interest in laser printer tracking dots and other techniques for embedding hidden messages in plain sight, also known as steganography. The development and deployment of these techniques comes out of a long-running secret relationship between governments and printer manufacturers in multiple countries. It sounds impossibly paranoid: every piece of paper that's printed by a color laser printer anywhere in the world is marked with invisible grids that make it possible for governments to trace when that exact page was printed and by what printer model. But in 2004, a Dutch journalist revealed the existence of the dots. In doing so, he sparked an ongoing arms race between privacy activists on the one hand and printer companies and intelligence agencies on the other.

Printing Money

For as long as commercial printers and copiers have existed, their manufacturers have been enlisted by US intelligence agencies to help track down whistleblowers. In 1971, the grassroots direct action group, Citizens' Commission to Investigate the FBI, raided a small FBI field office in Media, Pennsylvania and sent copies of the documents they found there—which first exposed the existence of COINTELPRO, the FBI's notorious program for sabotaging leftist organizing in the US—to news outlets across the country.

> *"The development and deployment of these techniques comes out of a long-running secret relationship between governments and printer manufacturers in multiple countries"*

At the FBI's request, Xerox helped the FBI determine that the Xerox 660 was the specific make and model of the copier used to disseminate the leak. Xerox also provided the FBI with a list of customers who leased the 660, but ultimately "decided at the very highest level" to not cooperate further with the bureau, according to Xerox PR manager Gerald A. Mulligan. Xerox's non-cooperation stonewalled the FBI's investigation. Even with over 200 agents on the case, the FBI never prosecuted anyone for the raid. Xerox's stand against intelligence agencies would be the last of its kind.

As personal computers, printers, and copiers became more readily available, intelligence agencies sought deeper inroads with printer technology companies in order to combat what they saw as a growing threat of currency counterfeiting. Japan was the first government to require traceable codes embedded in laser printouts and Fuji-Xerox, the joint Japanese-American printing venture, developed the first iteration of tracking technology—the yellow dots—in the mid-1980s. Other countries, including the US, quickly followed suit. Although there was no law in the US mandating that printer manufacturers incorporate MICs into their machines, industry insiders reported that the US government and intelligence agencies made clear to printer manufacturers that the lack of traceable markings could make it difficult to sell their products in the country. Soon, every major manufacturer implemented their own tracking codes. Not a single one revealed this "feature" to its customers.

By the 1990s, the desktop publishing revolution was well underway and central banks around the world took notice. In January 1993, central bankers and banknote printers from the G-10 countries convened a steering committee called the Special Study Group 2 (SSG-2) to assess the threat of laser printers to currency integrity. A 1995 SSG-2 memo describes a system for tracing counterfeit banknotes back to a specific copier. The SSG-2's next steps were to research "whether the technologies developed for copiers could also be used for scanner and computer systems"—with the goal being complete traceability for any given sheet of paper from any commercially available copier, scanner, or printer. A 1998 SSG-2 report remarks on the success of the tracing system and notes that "manufacturers will continue to provide assistance in identifying specific copiers at no additional cost."

Still, the details remained secret until 2004, when the Dutch journalist Wilbert de Vries broke the story of the tracking dots. In a brief article originally for the online IT publication Webwereld, de Vries cites anonymous sources with the Dutch Railway Police who confirm that they are using tracking dots to identify members of a counterfeit ticket printing gang. Privacy activists were appalled by the disclosure since the same technology that makes counterfeiting harder also subjects every single person who prints a piece of paper with a color printer — the vast majority of whom are not printing fake money or railway tickets — to non-consensual tracking. In the aftermath of de Vries's discovery, the Electronic Frontier Foundation (EFF) condemned the practice of using tracking dots and began collecting data on which manufacturers implemented them, including submitting FOIA requests to the US government regarding its involvement. Much of what we know about the history of printer steganography is due to their work.

> *"Not a single printer manufacturer revealed this 'feature' to its customers."*

EFF researchers were quickly overwhelmed by the sheer scale of the operation. They concluded that all major printer companies were implementing MICs in laser printers in some fashion. With the help of volunteers, they were able to decipher portions of the yellow dot tracking codes in use by a popular laser printer, the Xerox DocuColor. Each row of the grid was given a numerical value of a power of two — so the value of the bottom-most row would be one, the value of the next row up would be two, the next row up would be four, and so on up to sixty-four. Each column of the grid represented part of the printout's date or time or the printer's serial number. Specifically, columns 11 through 15 designated the printer's serial number. So if column 11 had a dot in row 32, plus a dot in row 16, the first digit of the serial would be 32 plus 16, or 48. The dot matrices produced by other printers looked completely different. Canon, for instance, arranges their dots in a spiral. Ultimately, Xerox abandoned their original grid pattern, making the EFF researchers' decoding tools obsolete. Newer printers presumably use a different system to encode metadata in the dots.

Following the Dutch revelations, printer manufacturers scrambled to justify their invasive and secretive technology. Xerox, for the first time, publicly admitted to using tracking codes and spun them as a positive, consumer-friendly feature to enhance the security of their printouts. As a result, they found themselves under the scrutiny of the EU Commissioner for Justice, Freedom, and Security, who stated in a 2008 official memo that such tracking "may give rise to the violation of fundamental human rights, namely the right to privacy and private life."

> *"One of our recent publications—Salty Wet, a visual essay on political resistance in Hong Kong—was covered in Xerox tracking codes."*

Disrupting the Dots

I run an independent press based in New York City. After learning of the pale yellow dots, I wanted to know if I could find them in the books, magazines, and chapbooks I had made over the years. When I scanned the pages into the computer and inverted the colors of the scans, I found that nearly all of them had the simple arrangement of dots that bound me up within a global surveillance apparatus. One of our recent publications—*Salty*

Wet, a visual essay on political resistance in Hong Kong—was covered in Xerox tracking codes. I wondered if there was any way to remove them, or if being trackable was simply the cost of printing in color.

In 2018, a group of researchers at the Technical University of Dresden made a breakthrough to that end when they released DEDA: the Dot Extraction, Decoding, and Anonymization toolkit. Analyzing over a thousand printouts from over a hundred printers, they developed an algorithm to detect and decode the tracking dots of four grid patterns that were used across eighteen manufacturers. Finally, there was a way for regular people to pry the dots out of the hands of corporations and intelligence agencies, and commandeer them for ourselves. The DEDA toolkit allows anyone to anonymize documents by removing the tracking dots at the software level, actively inhibiting the process and giving non-governmental entities the ability to disrupt their printer's surveillance mechanisms. It also enables users to hijack the MICs for their own purposes by creating user-defined secret patterns. Seemingly innocuous blank sheets of paper could now be used to convey information between any parties—not just central bankers, printer manufacturers, and intelligence agencies—so long as each party had access to the decoding key.

When I set up the toolkit on my laptop, I found that the anonymization and user-defined pattern tools worked right out of the box. It was thrilling to be able to manipulate the corporate tracking systems I had never consented to and create my own MIC. But when I attempted to decode the yellow dots on a piece of paper printed commercially from a Xerox PrimeLink C9065, DEDA came up empty. I tried another piece of paper from a smaller Xerox DocuColor printer and, again, DEDA found nothing, even after I had manually confirmed with a UV light that the page was marked with tracking codes.

Perhaps the toolkit is a victim of its own success. At the time, DEDA was a revelation. But security is not a steady state. The fanfare from the cybersecurity press that accompanied the toolkit's release may have caught the eye of printer companies and intelligence agencies, and maybe they tweaked the encoding schemes they were using. It is also possible that the printers I tested use a MIC that hasn't yet been catalogued by DEDA. According to the EFF, engineers employed by major manufacturers have hinted at the existence of a new generation of tracking mechanisms. It's been posited by researchers that tiny discrepancies in the spacing between words or even the kerning of letters could be used to encode information. But little is known about these alternative tracking measures aside from vague warnings by industry insiders.

The rationale behind building surveillance mechanisms into laser printers laid the foundation for more far-reaching forms of surveillance that we encounter in our devices today. The modern internet is full of invisible tracking mechanisms that, like MICs, are marketed as beneficial at best and harmless at worst—as long as you have nothing to hide. Compared with the sprawling digital behemoth of the web, yellow tracking dots may seem trivial. But the efforts by intelligence agencies to keep tabs on printed documents are a grave, if obscure, threat to our privacy. And the history of those efforts reminds us that what might at first sound like a conspiracy theory is actually true: that in the name of preventing crime, government and industry collude in secret to track us all.

Mitch Anzuoni is an editor and writer based in New York City. They help run Inpatient Press, an independent press dedicated to esoteric and experimental works.

My Blue Window

An Interview with American Artist

American Artist makes installations and sculptures about the foundational anti-Blackness in the networks and interfaces that shape our lives: the cell phone, the graphical user interface (GUI, or "gooey"), the #BlueLivesMatter movement. In addition to their writing, some of their most visceral critiques are visual. Shiny black liquid pools in front of an asphalt computer in an otherwise stark, white room. A black box, for baling hay by hand or maybe for storing servers, sits empty. In a moment when multiple crises loom, technology is often cast as humanity's path to security. American Artist's work looks at how security and humanity have always been conditioned on race. We caught up with them to hear more about their work.

In a few magazine interviews, you've mentioned that your exhibition, "I'm Blue (If I Was [redacted] I Would Die)" is inspired by Tiana Reid and Nijah Cunningham's essay, "Blue Life." I was wondering if you could say a little more about the idea of blue life and the significance of the essay to you.

The essay was my first knowledge of blue life. The authors, Reid and Cunningham, describe a bill that was put forth to further protect police officers—on top of the myriad laws that already protect them. They also explain why blue life is not a social or racial identity that anyone can inhabit and it is actually a metaphor of flesh for state power. This is all argued in relation to Blackness as a lived and politicized experience, which is once again dismissed by the proposal that blue life is another, significant social identity. When I read the essay, the language produced a literal blue figure for me. I wanted to contend with this visually, which is how the video in the exhibition, *Blue Life Seminar*, was constructed.

In the essay, the authors describe blue life as "constituted through the anticipation of violence and the projection of criminality. Blue life is not a personhood but rather a spectral legal identity that mimics vulnerability. Blue life is no more than a figuration. Blue life effectively obscures the violent operations of police power by attempting to give it flesh."

There's something about the classroom setting in your exhibition "I'm Blue (If I Was [redacted] I Would Die)" that hints at the ways blue life is indoctrinated into police officers. Can you talk a little bit about the role non-police (civilians) have in creating blue life? What is it about the idea of security or safety that some civilians have, where we agree to have our tax dollars siphoned to police officers?

The shape of the exhibition is a pedagogical space. Whether it implies a classroom, conference room, or police academy for viewers, it is nonetheless concerned with the relationship between power and ideology. #BlueLivesMatter was created for police, by police. In the exhibition, the "students" have their literal guards up, preventing them from hearing the contentious speech being broadcast from the front of the class. Within the desks are books that are, in some way, all concerned with maintaining this dangerous ideology.

Your question about how civilians construct blue life begs the following question: which members of society feel like the police work for them? Which ones feel comfortable to call the police on a public disturbance as if the police are their personal watchdogs? The rest of us don't feel like the police are interested in serving us or ensuring our safety. The entire concept of policing centers around an implied threat, which is us. For those who identify with this understanding of policing, the risk of arrest, excessive force, or even death increases when the police are called to a scene.

In New York, the notion of safety is very tightly linked to crime reduction and data-driven policing. This began with Bill Bratton's introduction of CompStat and the implementation of broken windows policing which created a precedent for the current wave of data-driven policing.

In your recent exhibition *My Blue Window* at the Queens Museum, you explore the ways software and predictive policing construct criminality. What do you see as the role of software in constructing blue life? Is there something intrinsic that you see in software that makes policing and software go hand in hand, or that explains the widespread adoption of policing software?

Definitely. For *My Blue Window* I wanted to continue framing the lens of policing as "blue" in place of any real lived social identity. The title implies a possessive relationship to the windshield of the police cruiser, which is the window that's being described. #BlueLivesMatter began with a hashtag. It is my opinion that the way people organize and understand identity in the time of social media is unique. The mental process we have become accustomed to, of projecting one's identity virtually, is one that allows a person to identify as blue, in a way that would have been less overt before the internet. This is how software helps construct blue life.

As far as software's role in constructing criminality, that's much deeper. Surveillance and policing softwares are a continuation of earlier forms of racialized capture, such as slave patrols, and records of physical descriptions of the enslaved and indentured used as a form of passport. Surveillance and policing derive from a time when Black people were merely property. These were tools used to secure the property. These technologies have changed form and become amorphous, but they can be traced to these practices directly.

I’m Blue (If I Was [redacted] I Would Die), 2019, Installation View (Koenig & Clinton, Brooklyn, NY). Photo: Jeffrey Sturges.

I’m Blue (If I Was [redacted] I Would Die), 2019, Installation View (Koenig & Clinton, Brooklyn, NY). Photo: Jeffrey Sturges.

My Blue Window, 2019, Installation View
(Queens Museum, Corona, NY). Photo: Hai Zhang.

Opposite page:
I'm Blue (If I Was [redacted] I Would Die), 2019, Installation View
(Koenig & Clinton, Brooklyn, NY). Photo: Jeffrey Sturges.

My Blue Window, 2019, Installation View
(Queens Museum, Corona, NY). Photo: Hai Zhang.

2015
BURGLAR
31

Still from 2015, 2019, Single-channel HD video, r.t. 21:56

06/22–06/28 2015 2014 PCT CHG
MURDER 5 7 -28.6%

The Party Decides

by Darren Loucaides

How the Five Star Movement hacked itself.

In late July 2017, a twenty-five-year-old hacker named Luigi Gubello emailed the staff of Rousseau to tell them he had found a security vulnerability in their software. Rousseau is the online platform that one of Italy's most popular political parties, the Movimento 5 Stelle, or Five Star Movement, uses to engage its supporters. Five Star claims to be at the forefront of digital democracy, allowing its members to vote online for everything from its policies to its political leaders. If Rousseau were compromised, Five Star's candidates for the Italian parliament could theoretically be selected by rogue actors, from lone hackers to foreign governments.

Gubello is a "white-hat" hacker, breaking into IT systems to highlight their vulnerabilities, not to exploit them. Gubello explained in his email to Rosseau staff that he had infiltrated the platform using SQL injection, a well-known type of vulnerability in which a data input field—username, say—is used to send commands to a database program, allowing a hacker to change or delete data entries, or download an entire database. It was the same sort of attack that played a major role in the famous hacking of Heartland Payment Systems in 2009, which caused hundreds of millions of dollars in losses to companies around the world, but it's also the sort of attack that a precocious high schooler might have launched in the early 2000s to boost her grades.

In his email to Rousseau, Gubello explained why the vulnerability was dangerous and how to resolve it. Shortly afterwards Gubello received a reply from Rousseau's staff thanking him, and claiming to have fixed the problem. The following month, in August 2017, Gubello, concerned about the continued vulnerability of Rousseau, went public with his findings in a blog post and urged the platform's users to change their passwords. "Hacker reveals: M5S's Rousseau site is not secure," declared *La Stampa*, the Italian newspaper.

At the time, Five Star was approaching one of its most consequential votes ever, to determine who would be the leader of the party and its de facto candidate for prime minister going into the following year's Italian general election. If the party triumphed then, this person could become the leader of the entire country.

On September 21, 2017, shortly after Five Star opened voting for its leadership election, the Rousseau servers stalled. Scores of members were unable to cast a ballot, and the party was forced to extend the voting until noon the following day. The party claimed this was due to the sheer numbers of people taking part, but of Five Star's 140,000 members, little more than 37,000 voted.

More than 80 percent of those voters selected a thirty-two-year-old politician named Luigi di Maio. It was a "world record," claimed a euphoric post on the party's blog: "the first leader of a political force chosen and voted for entirely on the Net." But then, just after voting finished, a malicious "black-hat" hacker named rogue0 began tweeting that they had compromised the election.

"Think of it like a house with many open windows," Gubello said of Rousseau when we spoke recently. "I saw only one open window"—the vulnerability he used to deploy his SQL injection attack—"and they closed it. But rogue0 found another one."

Open Windows

Among the tweets that rogue0 sent that night were screenshots of multiple votes he had generated for Luigi Di Maio, the winner of Five Star's leadership contest. The party claimed that every vote had been authenticated using text message confirmations, but rogue0 poured scorn on this. "Rest assured, Luigi Di Maio has already won, dozens of my certified votes assure you," the hacker tweeted. Even after the warning of Gubello, Rousseau—and Five Star's purported digital direct democracy—had been left exposed.

Another series of screenshots rogue0 posted appear to show the hacker signing into the accounts of various senior members of Five Star. In the subsequent hours and days, rogue0 also published some of the login details he had scraped from the Rousseau database; several of Five Star's leaders and digital operations staff had laughably basic passwords, such as their birthdays. "Seriously I can continue all night long this way," rogue0 tweeted at one point. Eventually, the hacker dumped the entire database into a publicly accessible pastebin.

Gubello suspects that rogue0 accessed the login details using another SQL injection attack that swiped an admin password from the database. Fabrizio Carimati, an Italian internet security expert, agreed. “In reality, Rousseau’s website was simply badly written, so everyone suspects that rogue0 used basic techniques,” Carimati told me. He added that the chances of the hack being a SQL injection attack were “99 percent.”

> *“He was a fan of Genghis Khan’s horseback couriers and Benito Mussolini’s radio broadcasts, and foresaw a future a few generations hence in which a total war would annihilate billions of people, leaving the remnant to govern itself by means of a worldwide internet democracy.”*

Rousseau was vulnerable in part because it was built using a proprietary content management system based on a version of the Movable Type software released in 2008. The upshot was that Rousseau had to be maintained in-house, and was only as secure as its best programmers could make it.

Other digital parties use more robust systems. Spain’s left-wing Podemos utilizes an open source and blockchain-based system for its online voting which claims to be tamper-proof; it also allows for external verification, which Five Star didn't have in place until 2018. The Pirate parties of Northern Europe, which, broadly speaking, advocate direct democracy and internet freedom, use open-source software such as LiquidFeedback and Loomio, which can be security tested and improved by almost anyone; similarly, organizations such as Occupy Wall Street and city governments in Reykjavik, Barcelona, Madrid, and Paris have adopted open-source voting tools.

It's difficult to say exactly why Five Star relied on old, self-maintained software. Perhaps it was simply a legacy system with high switching costs. But its proprietary nature also fit with the general culture of the people running the party. Though Five Star has presented itself as a populist movement empowered by digital tools, the reality is that Rousseau is controlled by a private firm, Casaleggio Associates, which operates with a great deal of secrecy and has exerted a significant amount of control over the movement.

A Sense of Community

The Five Star Movement was founded by Beppe Grillo, a comedian famous across Italy for his pugnacious humour, and an internet entrepreneur named Gianroberto Casaleggio—the founder of Casaleggio Associates. For more than a decade, Grillo has been the movement's figurehead. But until his death from brain cancer in 2016, Casaleggio, who was relatively unknown among most Italians, was the movement's most powerful figure.

Casaleggio had some intensely bizarre political beliefs that seemed to verge at once on the paranoid and the utopian. He was a fan of Genghis Khan's horseback couriers and Benito Mussolini's radio broadcasts, and foresaw a future a few generations hence in which a total war would annihilate billions of people, leaving the remnant to govern itself by means of a worldwide internet democracy.

But Casaleggio also had a more calculated side. Since the 1990s, he had sold businesses digital tools for managing employee sentiment. These included web-based forums in which workers could float new ideas and air grievances, all while being monitored by their bosses. Casaleggio would then help employers intervene strategically in the online discussion, promoting certain ideas while tamping down dissent. It was this approach—offering users a sense of community and self-determination while quietly shaping the discourse from above—that Casaleggio later instituted in the realm of electoral politics, first through a wildly popular blog he built for Grillo in the 2000s and later with Rousseau, which launched shortly after his death in 2016.

> *"Far from making the vote more democratic, digital tools helped the party's leaders engineer their preferred outcome."*

The 2017 leadership election that rogue0 hacked was a representative instance of Casaleggio's approach. Five Star initially appeared to many observers to be a left-leaning populist movement—its first policy positions included advocating renewable energy, revitalized public transport, and a universal basic income—that railed against the real and perceived corruptions of Italy's political establishment. But at heart both Grillo and

Casaleggio had more right-wing sympathies. Casaleggio's son Davide, who took over Casaleggio Associates after his father's death, shares these sympathies. His and Grillo's preferred candidate in the 2017 election was Luigi Di Maio, the son of an Italian neo-fascist.

Di Maio was already well known in the movement but, in the weeks leading up to the vote, Casaleggio Associates used Five Star's blog to praise Di Maio repeatedly. The result was more like a confirmatory ballot than a plebiscite: members were essentially rubber-stamping the decision of the party bosses. "The election of Di Maio can be described as a 'show election' in the sense that there was not real competition," the scholar Paolo Gerbaudo, whose book *The Digital Party* examines digital democracies around the world, told me. "All the candidates that could have stood a chance did not participate in the election, and the results were widely anticipated." Far from making the vote more democratic, digital tools helped the party's leaders engineer their preferred outcome.

Business as Usual

In the wake of Five Star's 2017 leadership election, and the hacks by Luigi Gubello and rogue0, the Italian Data Protection Authority conducted an investigation into Casaleggio Associates for breaking data privacy laws. The hacks had revealed that the company was collecting an extraordinary amount of members' personal data, along with their voting records, and combining this with data gleaned from members' social media accounts. It's the sort of information that could potentially be used to create highly targeted messaging for every member of the Five Star party, thus deepening Casaleggio Associates' centralized control over the ostensibly popular movement.

Around the same time as that investigation was underway, Five Star took legal action against Gubello, whom party leaders now accused of being rogue0. In January 2018, Italian police tracked down Gubello at his girlfriend's house in Trieste and examined his phone and computer. Meanwhile, Five Star claimed Rousseau's vulnerabilities had been addressed, and described the platform as a "fortress."

But in February, just a month before the 2018 Italian general election, rogue0 struck again, showing that they could still access the Rousseau database and take control of administrator accounts. In March, Five Star won a plurality of votes in the general election, and formed a coalition government with a nativist right-wing party called Lega, or the League. Di Maio was made the Minister of Labor and Social Policies, the Minister of Economic Development, and the Deputy Prime Minister. (He now serves as the government's Minister of Foreign Affairs.)

> *"The same technologies that are supposedly meant to empower members can instead provide vectors of control for party elites."*

Rogue0's identity has not been revealed, though the journalist Jacopo Iacoboni has perhaps come closest to discovering it. He has had a handful of conversations with the hacker via direct messages on Twitter. Iacoboni says it was rogue0 who reached out to him. "I never knew his real identity," Iacoboni told me, assuming the hacker was one person, and a man. "But he seemed to me someone who knew people at Casaleggio Associates very well—this was my impression after our Q&A."

Today, Rousseau's Movable Type content management system is gone. User passwords are now allowed to be longer than the previous requirement of just eight characters. "In fact, following orders by our [Data] Protection Authority, the Rousseau website was practically rewritten from scratch," noted security expert Fabrizio Carimati. But it's not clear exactly what's replaced Five Star's old systems. "Casaleggio now claims that the platform has been completely redone," Iacoboni said, before noting that this is impossible to verify.

But the vulnerabilities go deeper than software. Luigi Gubello believes that for voters to put their faith in digital democracy, the systems being used need to be transparent. Yet algorithms, coding, cryptography—the building blocks behind a digital democracy—require expertise to fully grasp. "People must be able to trust the election process," Gubello insists. "Can you trust it if you cannot understand and really check it?"

This is part of the paradox that seems to be at the core of all digital democracy movements: the same technologies that are supposedly meant to empower members can instead provide vectors of control for party elites. Even Podemos in Spain, with its populist politics and sophisticated digital transparency measures, mainly holds confirmatory ballots approving decisions already made by the leadership, as Paulo Gerbaudo notes.

Both Five Star and its Rousseau platform are less celebrated than they once were. The party has crumbled in opinion polls and in January 2020, Luigi Di Maio resigned as leader amid infighting and defections. For the time being, he has been replaced by a caretaker leader. Di Maio's resignation raises the possibility of another leadership contest conducted on Rousseau in the coming months. But the shine has gone from the party's digital democracy ambitions. "We hear little of Rousseau now, it's no longer really news," Carimati said. "That's because Five Star is one big crisis. And Rousseau is seen as increasingly irrelevant."

Darren Loucaides is a writer based in Barcelona and London.

"Free" as in Smash the Surveillance State

Alison Macrina on the Library Freedom Project and Tor Browser

There are a thousand reasons to feel sentimental about public libraries. Perhaps the most obvious is that they're freely accessible repositories of vast amounts of human knowledge. But librarian Alison Macrina gives us yet another reason: libraries can serve as the front line of a community response to mass surveillance. After all, there are thousands of them across the US. They have chairs and air conditioning and public computers. And, most critically, they have librarians.

In 2013, Macrina started teaching digital privacy workshops at the Watertown Free Public Library outside of Boston. In 2014, she founded the Library Freedom Project to expand that work to other libraries around the country. As part of her workshops, she has taught library patrons about Tor Browser, which looks similar to Google Chrome or Firefox but protects users' privacy and anonymity under the hood by proxying web traffic through multiple servers along the way to its destination. While Macrina was building the Library Freedom Project, she also led Tor's community team — work that took her around the world to conduct user research with activists who needed to protect their identities online.

Lately, she's been running a new initiative out of the Library Freedom Project: the Library Freedom Institute, which scales up her work by training librarians around the country to run workshops similar to the ones she used to run herself. We sat down with her to learn about privacy, power, and why librarians can save the world.

You founded the Library Freedom Project and you're also a core contributor to the Tor Project. Which came first for you?

Library Freedom Project (LFP) came first. I was working at a library outside of Boston in Watertown, Massachusetts. Through a confluence of forces, including living through the Boston Marathon bombing and the police militarization around that, plus the Snowden revelations just a couple of months later, plus the nascent but growing Black Lives Matter movement... all of it made me think about technology and policing and privacy in a way that I never had before.

I started LFP in the summer of 2014. At first it just involved teaching people about privacy in my library, doing some introductory community events. When there turned out to be a lot of interest, I cold emailed the people at Tor saying, "Hey, I'm a librarian working on privacy, and I'm teaching people about Tor. Can I be in touch with your outreach people?" And they responded, "We don't have any outreach people. Do you want to be our outreach people?" That's how the Tor connection happened.

I read that the Library Freedom Project's pilot project was setting up a Tor relay in a public library. What is a Tor relay?

The Tor network is made up of nodes run by volunteers. When someone uses the Tor Browser from their own computer, their web traffic gets bounced to all those different volunteer nodes so that someone looking from outside can't trace that person's web traffic back to their computer at their house. A Tor relay is one of those nodes; it's a computer that's configured to just forward traffic all day to other nodes of the Tor network.

We did set up a Tor relay in a library in New Hampshire around the summer of 2015. It wasn't quite LFP's pilot; I had already been running LFP for about a year at that point. But a lot of people first learned about us because the relay got the attention of the Department of Homeland Security (DHS), and that whole situation got some press coverage.

How did the Tor relay project come about? And what was your pitch to that library?

I had been traveling around the country doing these direct trainings, building relationships with librarians and helping them understand the surveillance problem from justice-focused angles. I was focusing on strategies that they could employ, both in their direct work with patrons, and then also in a bigger, advocacy sense—macro-level stuff they could do out in their communities.

The Tor relay project came about because I started thinking about how libraries are in a strong position to run privacy infrastructure. And Tor relays were the perfect example. In order to run one of these things, all you need is a little bit of bandwidth. It also helps if you are an institution rather than an individual, because law enforcement might be more likely to hassle an individual, especially if they run one of the exit relays, which is identifiable on the network.

I thought, "Libraries are ideal for this." Fortunately, because I had built all these relationships, I knew who the other hardcore privacy-minded librarians were. One of them was my friend Chuck McAndrew at this library in New Hampshire. He was already a Linux guy and he was as evangelical about privacy as I was. He was 100 percent on board with being the first library to set up a Tor relay. The conversation asking him to do it was quite simple. I was like, "Maybe you want to do this?" And he was like, "Yeah." So it was easy.

> *"The Tor relay project came about because I started thinking about how libraries are in a strong position to run privacy infrastructure."*

The Tor Project has its critics. Did you get any pushback?

We set the relay up and promoted it. It was a cute little story that ran in *Ars Technica*. And then the library contacted me a few weeks later and said, "We've gotten this email from our local police department who've been contacted by DHS. They said that we're not allowed to have the relay and that we have to turn it off." The library was understandably frightened.

I remember being pretty taken aback when I first read the email. But then I thought, "Well, this is the best thing that's ever happened to us." At that time, people weren't thinking about DHS in the way that they are now—I mean, certainly immigration activists were, but the general public was not. You might recall that before Trump, DHS was seen as the ugly stepchild of the intelligence community. They're a Bush-era construction. People were like, "What do they even do?" And the "homeland" thing in their name is just creepy. So of all the federal law enforcement agencies that could have gone after us, they were the best because they're total yahoos.

We rallied around the library and got a lot of support from the ACLU and the Electronic Frontier Foundation and the local community. The libertarian separatist community that lives in New Hampshire even showed up. It was a great demonstration of support for the library and privacy, but also a broader rejection of that kind of law enforcement interference.

We had a big public hearing, and you can imagine what a library board meeting on a Tuesday night in rural New Hampshire is normally like. But that night, it was standing room only. A bunch of press came. Meanwhile, no one from DHS showed up. They knew they had no legal grounds to ask the library to remove the Tor relay.

So anyway, the relay got turned back on and it was a lovely story. That was the first big thing that happened for LFP.

That's wild. How did DHS find out about the relay? Were they monitoring the library's internet traffic or social media or what?

I guess they read Cyrus Farivar? He was the journalist who wrote the article about us for *Ars Technica*. Someone later did a FOIA request for the DHS emails and found that an agent linked that article and then said something funny about me. That was basically it.

Privacy School

When did the Library Freedom Institute (LFI) happen?

People were super into the LFP trainings. I was getting invited to speak to a lot of librarians. It became apparent to me pretty quickly that I couldn't scale the project in the way that I needed to without bringing other people in somehow.

> *"Of all the federal law enforcement agencies that could have gone after us, they were the best because they're total yahoos."*

It was also clear to me from the beginning that we needed to have a community response to surveillance. I really wanted to build a community of practice that could meet that challenge, so I started thinking about what that could look like. I developed a blueprint for a six-month-long train-the-trainers workshop where librarians located in different places could get together online every week and learn about this stuff. I got grant funding, and our first LFI cohort of thirteen librarians started in June 2018.

We're currently experimenting with different timeframes, but LFI workshops currently run for four to six months. Librarians apply to get into it. It's totally funded by the Institute of Museum and Library Services (IMLS), so it's free to attend. We spend each week online together thinking about a different aspect of the surveillance problem. To date, there are forty-three librarians that have finished the program, and another twenty librarians will begin the course in March 2020.

Where are these librarians joining from?

We have librarians in twenty-one states. In this last LFI cohort, we had three people in Georgia, somebody in West Virginia, a couple of New Yorkers, a few in the Bay Area, one at Los Angeles Public Library, Boston Public Library, Hennepin County Public Library in Minneapolis. All over the US. For now, we're restricted to the US since our funding is through IMLS, a US federal agency.

I make sure there are a lot of folks from rural libraries and Southern libraries since those libraries tend to be under-resourced and far from in-person training opportunities. We also prioritize public librarians because they tend to interact with the public more than academic librarians, and also because public librarians generally don't have the professional development funding to make something like LFI happen on their own.

Do they come representing their patrons or are they also curious for themselves?

A little of both. In order to be accepted into the program, there's a short application and it's basically a free-form questionnaire about what they see in their communities and what their personal experiences are that make them a fit for LFI. Sometimes their community is representative of who they are.

One of our librarians from the last cohort is Anishinaabekwe of the Sault Tribe of Chippewa Indians. They work in a non-Native community now, but their whole objective was to think about how surveillance affects Indigenous people. How, for example, Indigenous women are more subject to certain kinds of both domestic violence and state violence, and how surveillance intersects with that.

To talk about surveillance online, you have to talk about how the internet works, and it seems like a fine line between being too high-level and too technically nitty-gritty. How do you teach about the internet?

Most of the time, what we're teaching is fairly 101-level. In terms of getting into the structure of the internet, we try to avoid getting in the weeds just because most people, whether they're the librarians who we're training for the first time or the patrons who our graduates go on to train, come into a privacy-focused educational setting feeling hopelessness and resignation. Often, they're dealing with a problem that is happening to them right now, like stalkerware or having had their identity stolen. They're already freaked out. Most people think they don't have a good sense of how technology works. They already feel stupid asking questions about it.

In LFI, we work with a trainer named Mallory Hanora, who comes from the world of prison abolition and teaches us how to create transformative workshops. Their framework is about incorporating the experiences of the people in the room, being as nonhierarchical as possible, meeting people where they are—all these sorts of things to create an environment that people can actually learn in.

Then, LFI graduates are able to bring that kind of framework forward in their workshops with their own patrons. Usually when they teach, no matter what the focus of the lesson is, people will say during the question and answer period at the end things like, "I started using this password manager. Is this good?" Or, "My nephew told me about this thing I can download to block ads. What do you think about it?" That is fascinating to me because it confirms to me that people do care about their privacy. They are already taking steps to mitigate or manage what is revealed about them on the internet. They haven't developed a nuanced threat model for their own unique situation, but most people are already doing something. In those moments, LFI librarians have the information and facilitation tools to help patrons build on what they already know.

User Testing in Uganda

I want to make sure we also talk about Tor. What does your work on the community team look like?

I was the community team lead for two years and I'm still a contributor, though not at the level I used to be as Library Freedom Project has taken up more of my time. The community team works on outreach—teaching people about Tor and getting more of an understanding of who it's for and what it does. LFP really fits into that work. But beyond LFP, Tor is a global community, so our outreach is global.

The interesting thing about Tor software development and usability research is that we don't do the surveillance stuff that some software does to make sure that it's usable for you. We don't track your clicks or how your mouse moves around the page; we're not doing A/B testing and getting analytics back. So how are we supposed to know how well Tor is working for you? For years, we got that feedback directly from the most technical people because that was who was using Tor. But the problem with that method was that it meant we were designing for the most technical people, and if only the most technical people are using Tor, then Tor doesn't actually work.

A couple of years ago, we started focusing on making the Tor Browser usable for people doing social justice movement work—water rights, reproductive justice, queer and trans issues—around the world. We wanted to focus on members of these movements in the Global South for a few reasons. One is that we didn't want to have the kind of free software project that was all white dudes in the US and Western Europe, which is what most of those projects are and, to be honest, what ours kind of was. We also thought it was important to understand the context of using Tor in places where the internet is much slower and more expensive—places where there are all kinds of environmental factors that make using it harder, even if you know how to use it technically.

Our teams started traveling to meet with community organizers who were part of political movements in different parts of the world. We spent a lot of time learning about their contexts, what kinds of work they do, and who their adversaries are. Then, we showed them the Tor Browser, conducted some user tests, and asked for feedback. That's how Tor usability work has been happening for some time now.

Can you give an example?

In 2018 I traveled to Uganda with Antonela Debiasi, the head of Tor's usability team. We visited two cities: Kampala, the capital, and a smaller city called Hoima. Hoima is in what's known as the Albertine region of Uganda, after Lake Albert. They discovered a lot of oil in the Albertine region around fifteen years ago, and US, Chinese, and Israeli companies have been scrambling to extract the oil in that region since then.

Antonela and I met with a number of people in Hoima, mainly environmental activists. They were mostly doing journalism, writing and publishing anonymously to try to bring international attention to the environmental effects of large-scale oil extraction that they were experiencing. We asked them who their adversaries were and what they knew about the companies operating there. They had some anecdotes about people being followed and weird things happening with their phones.

In that situation, we were able to bring knowledge about the surveillance capabilities of these big companies. We talked about different kinds of monitoring that may have been happening, like Stingrays for intercepting phone calls. Of course, Tor can only solve a piece of those issues, but it can really help with anonymous research and publishing, especially if you teach people about what its limitations are. And it's like that with every different group: we learn about their context, we think about what Tor can do in that specific context, and we share information about other ways people can protect themselves. The process was similar when we went to Kampala, although the activists we met were focused on different issues. There are a lot of refugees there from Sudan and South Sudan, so people were working with them. And we met with reproductive and gay rights activists.

> *"If it's not safe for them to download the files directly—because evidence of the download would be visible to their ISP—then we can sideload them using a USB stick."*

In both cities, we were meeting with a combination of different people who knew each other through their own networks. I think that is fairly typical. In some cases, we try to go back again and meet with people a second time. But, in general, we learn all these things and then we take that information back to the developers and incorporate what we learn into the Tor Browser.

What were the meetings like?

In general, our workshops with activists are fairly straightforward. We get together in a classroom and learn about their threat models—what kind of work they're doing, who their adversaries are, what kind of work they would be doing if they felt less threatened by those adversaries. Since we work with people who are all familiar with each other through social and political networks, they tend to feel comfortable sharing these details in a group. After we get a better sense of their work, we give them a hands-on training on using Tor. We have them use their own phones and laptops, and we walk them through the download process while explaining what Tor does and doesn't do. If it's not safe for them to download the files directly—because evidence of the download would be visible to their ISP—then we can sideload them using a USB stick. We troubleshoot any issues that come up, and we learn about unique local challenges like data limitations, slow internet, or electrical outages. We think about the specific adversaries that these activists are facing and what kinds of capabilities those adversaries are likely to have.

For example, from what we know about the companies operating in the oil market in Uganda, we can reasonably assume that they're employing some security firm like NSO Group to spy on the activists. [*Eds.: NSO Group is an Israeli spyware company that sells surveillance software to governments.*] And we know that the Ugandan government is monitoring them. Knowing those specifics helps us explain to activists how Tor can help. We also spend time conducting user tests, asking activists to complete a series of tasks using Tor Browser, and documenting what was difficult.

Beyond the fairly unique threat models that people are working with, it's helpful to know how internet infrastructure impacts Tor use and performance in different places. For example, using Tor in San Francisco is slower than using Chrome in San Francisco because, in order to anonymize your traffic, Tor bounces your web requests to relays all over the world before it returns the website you want. So we figured that a slower internet connection would make Tor even slower. How much slower? In Uganda, it turns out: not much! The electricity goes in and out, so you get disconnected, but that impacts every browser the same way. It's also helpful to know how much data Tor is using and how much mobile data plans cost; a lot of the people we met there are primarily browsing on their phones and we don't want people to have to use up their whole data plans. We learned about censorship by internet service providers there and which anti-censorship mechanisms that are built into Tor will work in those contexts. There's all sorts of feedback that's coming to us in those meetings. We take all this information and write up a report for the developers and the user experience team so that they can address any challenges that arose.

Socially Necessary Library Time

You've said that the discussion about privacy is really a discussion about power. The fact that Tor solves a similar privacy problem for environmental activists in Uganda and also for librarians in the US seems to highlight that.

Absolutely. That's the common thread in all of this, and that's my approach with Library Freedom Project. To me, privacy is the Trojan horse. Then when I get in the building, I'm like, "Alright, let's actually talk about capitalism." Privacy is important, but what we're really talking about is the newest frontier of exploitation by capital.

With both Tor and the Library Freedom Project, you're making libraries places that are anti-capitalist not just because

they're free, but also because data harvesting won't work in there. You're making a Faraday cage around libraries.

I love that. I do think of libraries as one of the most socialist institutions that we have in the US in 2020: they are funded with public money. They are not means-tested. They're everywhere. You talk to the average librarian about why they got into the profession and there is a real love for the public. You can go into a library and you don't even have to use it for its explicit purpose of looking at books or magazines or computers. You can just go there and be, even if you're the kind of person that late capitalism has decided is not worthwhile. To the library, you are. And this is a radical idea. Libraries' doors are still open, even as their budgets get cut. They're special spaces and I want to help them expand what they're capable of doing.

> *"Privacy is the Trojan horse. Then when I get in the building, I'm like, 'Alright, let's actually talk about capitalism.'"*

At the same time, right now, seven of the top ten companies by market capitalization are tech companies. Seven out of ten are using data that they take from us, without our consent, to create their products. That is part of our labor power: those products are made with our emotional labor, our mental labor. Privacy is a way to reclaim our labor power. I want people to think about those relationships.

And, yeah, I also want people to not get their identities stolen. All of the more concrete problems are still important to me, especially when you think about who is subject to them—it's poor people and elderly people and people who don't have power. But with all of this work, I'm really trying to force a conversation about who controls the internet and what that means for our lives.

Hacking Security

by Matt Goerzen and Gabriella Coleman

Hackers helped invent the field of computer security. Their ideas can help us revamp it for a new era.

In August 2019, Alex Stamos took the stage at the USENIX Security Symposium to deliver the event's keynote presentation. "I'm a little nervous today," he began. "This is not my normal crowd."

Indeed, as a Silicon Valley insider, the former chief security officer of Facebook was not the sort of academic computer security expert who typically delivers the event's opening talk. After cracking a couple of self-deprecating jokes, Stamos offered what he saw as the reason he was invited to speak: technical security specialists have not kept up with the changing nature of threats stemming from the imbrication of information technology with every aspect of our lives. "It turns out the vast, vast majority of the human harm—of the people who are actually suffering in their lives because of tech—are suffering because of what we term in the industry 'abuse,'" he said. "Abuse is the technically correct use of the products we build, to cause harm."

Stamos illustrated his point with a simple diagram. It consisted of a triangle, with a small portion at the top labeled "InfoSec"—short for "information security"—and a large base at the bottom labeled "Abuse." "The top of the triangle is basically everything that all of us in the room have dedicated our lives to," he explained. But the bottom of the triangle—"abuse," in Stamos's terminology—is a far more prevalent form of technological exploitation. Stamos suggested that traditional security experts, himself included, are ill-prepared to deal with this class of threats. The computer security industry occupies itself with identifying and fixing technical flaws, the kind that hackers might use to gain unauthorized access to the backend of a system. Such work is important, but it leaves aside whole categories of more common dangers to people online.

These dangers include child pornography, trolling, and vengeful ex-boyfriends who hound their former partners, just to name a few. Most recently, disinformation and Russian political meddling to fuel polarization have become prominent in our daily news. Often, the abusability of certain features stems from engineering decisions that optimize for profit. Ad-driven platforms are designed to maximize engagement metrics so they can sell more eyeballs to advertisers. This means the algorithm prioritizes clickable and shareable content—even if that content is harmful. But this sort of abuse can be harder to address than the vulnerabilities that come from technical flaws.

Consider, for example, the way online content spread in the wake of the white supremacist terror attack in Christchurch, New Zealand. In March 2019, a gunman killed fifty-one people and injured fifty at the city's Al Noor mosque. He livestreamed the first seventeen minutes of the attack on Facebook, sharing it on an 8chan imageboard favored by white supremacists. He also took steps to ensure his livestream would evade detection and removal. By inserting arbitrary spaces to break the link to his stream in his 8chan post, he forced people to manually reconstitute the URL rather than click through the link directly—a move that could alert Facebook to the source of the traffic. We can also speculate about other flaws he may have exploited, such as the way he talked for a period of time before initiating the attack, perhaps knowing that any content moderators reviewing the stream would then "snooze" it; or the way his head-mounted camera angle simulated the aesthetic of a first-person shooter video game, in a manner that may have been sufficient to trick a machine-learning analysis.

That the livestream went undetected for so long represented a colossal failure on the part of Facebook—and one that the traditional computer security industry would have little to say about. The problem wasn't just technical, but social. It involved countless complex interactions within a stack of sociotechnical workings—"trust and safety" specialists, content moderators, machine learning algorithms, and so on. The result was disastrous: a white supremacist was able to broadcast a massacre around the world, his supporters were able to redistribute the video across the web, and the images he posted to various social media accounts—littered with Google-able keywords that would lead the curious to pro-white supremacist content—were amplified by subsequent media reports.

How can we begin to combat this sort of phenomenon? Stamos urged his audience to start tackling abusability with an interdisciplinary approach: to increase the representation of communities affected by abuse and to create more institutions—like the one he now heads at Stanford University—that enable technical security specialists to collaborate with social scientists.

These are good ideas. Bringing in outside voices and experts to revamp the field of computer security is important. But it's also important to recognize the limitations of the field itself. It's not enough to broaden the scope of what security experts work on, or who a security expert even is. We also need to recognize how outsiders can identify issues, put them on the public agenda, and in doing so reorient our notions of what security is.

The history of the computer security industry itself can help. To refocus security, we first need to revisit the story of how it came to be—how practices to address technical vulnerabilities originally emerged. Those practices sometimes provided valuable consumer protections, but they were also unsatisfactory to many. The computer security industry came to serve a narrow band of people and interests, creating norms about what kinds of security mattered, and for whom. Yet other possibilities were available, generated by quieter, countervailing strains that proposed different answers to the questions of whose security was worth taking seriously, and what it took to make them feel secure. By mapping out these alternative paths, we can better understand what a traditional technical security paradigm can usefully offer, and what must be discarded, modified, or challenged as we find new ways to face new threats.

An Industry is Born

Today, the computer security industry and its associated institutions are firmly established. Along with dozens of high-profile companies dedicated to diagnosing, fixing, and preventing breaches, many corporations hire specialized professionals to deal with a host of technical security issues. All of this activity adds up to a lot of money: analysts estimate the security industry to be worth about $170 billion.

This industry hasn't always existed, of course. It had to be built. And among the most important contributors were underground hackers who enjoyed breaking into systems and sharing what they found with their peers.

In the early days of computing, security largely revolved around control of physical access to shared machines. Even the idea of passwords was, for a time, controversial. While engineers developed computationally secure systems like Multics in the 1960s and 1970s, they were eclipsed by the popularity of Unix. The advent of computer networking revealed the shortsightedness of some of the design decisions baked into Unix, as the choices made in a pre-networked era left critical vulnerabilities for later hackers to exploit.

> *"To refocus security, we first need to revisit the story of how it originally came to be."*

Hackers gathered on dial-in electronic Bulletin Board Systems (BBSes) in the 1980s, swapping tips on how best to gain access to the different types of computers connecting to the nascent internet. Some joined underground secret societies, competing with each other for control of more computers. Others played cat-and-mouse games with systems administrators, who often lurked in the same BBSes, learning how to defend their users.

But in the mid-1990s, after a wave of high-profile arrests and the growing demonization of hackers in the media, some underground hackers began to shift their focus. Rather than hoarding their knowledge within select circles, they began to publicize it more widely. Upon discovering vulnerabilities, they would notify the company that made the software. Yet these vendors frequently reacted by dismissing the flaws as "theoretical," while telling their users that the only threat to their security came from the hackers themselves.

In response, certain hackers set out to "make the theoretical practical," as the tagline of the legendary hacker group the L0pht put it. Throughout the 1990s, they worked hard to force tech companies to take security seriously. Some put pressure on vendors by sharing the vulnerabilities on "full disclosure" mailing lists like Bugtraq and in hacker zines like *Phrack*. This move was risky—and controversial—since it potentially put exploits in the hands of malicious actors. But supporters of this approach hoped the risk would spur a rapid response from the vendor, while also empowering individual systems administrators to ensure their own organizations were adequately prepared.

Other hackers developed tools that made the security holes in popular software unignorable, by allowing even the non-technical to crack passwords or control users' machines remotely. Emblematic of this approach was a tool called Back Orifice. Developed by the Cult of the Dead Cow (cDc), it facilitated the remote control of Microsoft Windows machines—with or without the users' consent. Around the same time, the L0pht released a tool that gleaned passwords from Windows systems. The stated goal was to help systems administrators recover lost user passwords. But its demonstration of Microsoft's lax approach to password management was unmissable.

Such strategies helped create media buzz about bad security. The tagline of cDc, "Global domination through media saturation," made its mission clear: rather than simply hacking software systems, they were also hacking media narratives. "Was releasing Back Orifice to the public immoral? Microsoft would love for their customers to believe that we're the bad guys and that they—as vendors of a digital sieve—bear no responsibility whatever," wrote the cDc's Deth Veggie in a 1998 "morality alert," issued in response to a Microsoft press release. "We'll frame our culture and actions against theirs, and let the public determine which one of us looks better in black."

> *"Certain hackers set out to 'make the theoretical practical,' as the tagline of the legendary hacker group the L0pht put it."*

At first, however, these efforts didn't make much of a dent. Microsoft and other vendors responded by denying responsibility and vilifying the hackers. They insisted that their software would be secure if the hackers were held responsible—ignoring the obvious point that hackers didn't create the weaknesses they revealed.

But the hackers didn't only demonstrate insecurity from a distance. They began to remake themselves as trustworthy insiders as well. Some, like the L0pht frontman Peiter "Mudge" Zatko, started meeting with corporate executives and elected officials. They promoted a vision of ethical "grey-hat" and "white-hat" hackers who probed systems not for destructive purposes, but to gain insight into how best to secure them against malicious "black-hat" hackers.

Gradually, this message filtered into the mainstream. By the end of the 1990s, the dotcom boom was hitting its full stride.

The Clinton Administration began heeding warnings of a future "cyber Pearl Harbor" that could cripple critical network-connected infrastructure. Between a booming tech sector and rising concerns about a new kind of national security threat, the stakes of computer security suddenly started to seem a lot higher.

By the early 2000s, many companies were chomping at the bit to hire hackers. Even Microsoft changed its tune, recruiting from the community it once antagonized. Throughout the tech industry, hackers joined in-house security teams, led "penetration tests" to probe systems for vulnerabilities, and conducted forensic investigations to ensure successful attacks couldn't happen again. Hackers and vendors together crafted policies of "coordinated disclosure," ensuring vulnerabilities would only be disclosed after they were fixed—or if vendors failed to take their responsibilities seriously.

Many hackers were happy to take the lucrative jobs on offer. They could make a good salary doing the puzzle-solving research they loved, without the prospect of prison hanging over their heads. In 2000, the L0pht merged with @stake, a newly minted security firm that proudly touted its payroll of hackers—and which would go on to create a new generation of professional security experts, including Alex Stamos himself. After 9/11, several hackers began consulting for the US government and military. Mudge from the L0pht devoted himself to researching national security threats, joining a government contractor before ultimately taking a position as DARPA's cybersecurity research program manager. The barbarians at the gate were now inside the walls, helping to guard them.

Outside the Gates

But not all hackers went pro. Others remained fiercely independent. They revolted against what they saw as co-optation, and instead laid the groundwork for a variety of oppositional currents, such as "hacktivism," "anti-security," and "digital security," that challenged the notion that improving the security of machines, software, and networks is enough to really improve the security of those who use them—or whom they are used upon.

> *"By the early 2000s, many companies were chomping at the bit to hire hackers."*

First coined in 1995, the hacktivism label soon came to refer to a range of politically motivated hackerish pursuits, from technical support for activists to digital direct actions. For instance, the collective Electronic Disturbance Theater orchestrated acts of "electronic civil disobedience" to draw attention to the plight of the Zapatistas in the 1990s. They built a tool called FloodNet that, when used by multiple users, would "flood" a targeted website with traffic. The resulting "denial-of-service" attacks against Mexican and American government websites were used to spread awareness of the Zapatista cause. Many members of the hacking group cDc also took the hacktivist path, lending support to Chinese dissidents.

Other hackers took a different route, doubling down on their underground status. In the early 2000s, they coalesced around the banner of "anti-security." A heterogeneous movement, anti-security was full of conflicts and contradictions. But participants shared a basic skepticism about the notion that public awareness of vulnerabilities was necessarily a good thing. Embracing a technocratic, even aristocratic outlook, many in the anti-security movement argued that only underground hackers should possess knowledge of technical exploits. As anti-security failed to slow the security boom, however, some of its members adopted more drastic measures: they encouraged their peers to hack the white hats as part of "pr0j4kt m4yh3m." ("Project Mayem" was a reference to the movie *Fight Club,* which had just come out.) They were so successful that white-hat professionals began to see getting hacked as a rite of passage.

> ***"Security research was now being weaponized to promote forms of insecurity—helping governments crack down on dissidents, for example."***

Then, around 2010, the philosophy of anti-security and the tactics of hacktivism began to combine. One offshoot of the hacktivist group Anonymous paid direct homage to the anti-security movement, adopting the name AntiSec. The new crews exploited technical vulnerabilities in order to exfiltrate documents in the public interest and shame governments and corporations for their lack of transparency. Crucially, they didn't shame them for their failure to take security seriously. Rather, they targeted these entities because of *how* they deployed security technologies—namely, to extend the reach of corporate and state surveillance.

Important also was the work of groups like Citizen Lab. Created in 2001 by a University of Toronto researcher partly inspired by cDc, Citizen Lab enlisted computer security techniques such as auditing, threat modeling, and penetration testing in support of civil society—an approach sometimes called "digital security." Its research put abuses of power by states and corporations on the security agenda. After the organization conjectured that private security companies were helping governments commit human rights abuses by selling them hacking tools, a hacktivist named Phineas Fisher hacked and leaked documents and source code from the offending companies, making the theoretical practical and focusing public outrage.

Unlike in the 1990s, it wasn't enough for organizations to take technical security seriously. Security research was now being weaponized to promote forms of *insecurity*—helping governments crack down on dissidents, for example. Building on the earlier ideas of the anti-security movement, these watchdog hackers made clear that identifying vulnerabilities and passing the information along to the authorities was far from sufficient to improve everyone's security—in fact, it could be actively harmful. Security for some could mean insecurity for others.

Together, these alternative hacking communities challenged the idea that security was best served through the corporate and state securitization of communities and computer systems. They presented a different vision, where security meant something more than only technical considerations, and all people, not just the powerful, had a right to it.

Security for All

So what can this historical gloss teach us about how to confront the problem of abuse?

To flesh out some thoughts along these lines, let's return to the Christchurch shootings. In the aftermath of the attack, a debate emerged about the role of both Facebook and a company called Cloudflare in enabling the spread of the shooter's message. 8chan, the site where he shared his livestream, used Cloudflare to provide network redundancy in order to ensure it could handle heavy traffic loads. Viewed another way, 8chan relied on this service to protect itself from the kind of denial-of-service attacks first pioneered by hacktivists in the 1990s, and since adopted as a standard weapon of cyberwarfare. More than a few hacktivists—and plenty of non-technical people as well—wanted 8chan off the web. But so long as the site used Cloudflare, it was protected against denial-of-service attacks. So a growing chorus of individuals and organizations began calling for Cloudflare to sever its ties.

> *"In many cases the business model itself might be the foundational vulnerability."*

Questions about the nature of Facebook and Cloudflare's services and business model—as well as broader debates about corporate responsibility, free speech, and regulation—ricocheted throughout the public sphere, including in the hacker community. Following the attack, Alex Stamos joined security researcher Patrick Gray on the popular security podcast *Risky Business*. Gray insisted that Facebook remove white supremacist supporters and that Cloudflare, as a private company, exercise its right to drop 8chan as a client.

Stamos was hesitant, pointing out that Cloudflare had previously removed ISIS content only after being legally compelled by the government. He wanted to see companies embrace standards for determining the intentionality behind content, barring clear legal guidance or capacity. But for Gray, the moral case from the court of public opinion was clear enough. Echoing the 1990s-era hacker philosophy of accountability, Gray said, "I think we need to turn this to a discussion of brand protection, and if Facebook wants to be associated with stuff like that..." Both Gray and Stamos may have wanted the same outcome, but they saw different means to achieve it.

The case of Cloudflare clarifies the stakes of choosing different paradigms for security. At some point, technical questions about how vulnerabilities are identified and mitigated collide with questions about how technical security relates to other forms of security. When is a site like 8chan benefiting from technical security that enables its members to make other communities less secure? Drawing on the insights of hacktivists, anti-security hackers, and digital security advocates, we always have to ask who security works for—a sociotechnical line of thinking that computer security professionals, often keen to avoid anything that might be seen as political, may find uncomfortable.

We also have to foreground the role of the profit motive in making platforms prone to abuse. Data mining, corporate ownership of user data, targeted advertisements, design and policy decisions to maximize user engagement—these can all create opportunities for bad actors to cause harm. Put simply, in many cases the business model itself might be the foundational vulnerability. But such fundamental product and business issues are generally outside the scope of what technical security researchers working for these organizations are paid to identify and remediate.

After two more 8chan-linked shootings, Cloudflare ultimately caved to public pressure and cut ties with the site. But a million similar issues are raised on a smaller scale every day, where the question isn't whether to host a single site, but how to treat a particular piece of content, or a feature that allows that content to be promoted to a particular person. These issues can't be adequately addressed simply by tweaking interaction algorithms, removing "like" buttons, or developing better content moderation protocols.

A reassessment of what is involved in "security" is required. And, as with the 1990s hackers, this notion of security must be built, at least in part, by people who aren't afraid to pick a fight. What made an earlier generation of hackers so effective wasn't just their technical expertise but their willingness to antagonize the big software vendors. It was through their efforts that we now enjoy a modicum of technical security. Standards and protocols that protect consumers and citizens from harms like infrastructural sabotage, identity theft, and commercial fraud exist because hackers aggressively drew attention to corporate incompetence and demanded accountability.

But, as many of these same individuals entered industry in the 2000s, they lost their oppositional edge. To ensure security serves more than just corporate or state interests, this sort of adversarial spirit must be expanded. Stamos's call for the inclusion of sidelined expertise in computer security is important, but it needs to be accompanied by the adversarial work of activists, civil rights groups, and inquisitive tinkerers.

This work is currently being done by a range of groups, including Citizen Lab and other organizations engaging in digital security activism like Equality Labs and Access Now. This work can also take a variety of forms. In 2016, investigative journalism outfit ProPublica exposed how Facebook's targeted advertising algorithm can be used in "digital redlining" practices to exclude certain demographics from ads for housing. In 2019, computer scientist Jeanna Matthews and her interdisciplinary collaborators reverse-engineered proprietary software for DNA testing that is used in criminal trials, demonstrating its worrying propensity to deliver false positives. In 2020, Elizabeth Warren's presidential campaign deliberately lied in a Facebook ad in order to shame the company for refusing to fact-check political ads.

These are only a few examples. Taken together, however, they suggest the role that outsiders who "make the theoretical practical" can play in demanding a more people-centered definition of security—one that involves keeping everyone safe. After all, outsider pressure helped shape our ideas of technical security. Outsiders can reshape them.

Matt Goerzen studies sociotechnical security issues at the Data & Society Research Institute.

Gabriella Coleman holds the Wolfe Chair in Scientific and Technological Literacy at McGill University.

Attacking Agriculture

by Rian Wanstreet

Agricultural production increasingly relies on sensors and internet-connected devices that are rife with security vulnerabilities.

On a rectangle of land the size of a baseball field three hours northwest of London, a project called Hands Free Hectare plants and harvests wheat with almost no human intervention. For the last three seasons, an autonomous tractor has worked its way up and down the plot, sowing the wheat and spraying it with liquid fertilizer, herbicides, and fungicides. A small robot collects soil samples and a drone takes thermal photos of the crop from above to monitor for inconsistencies. When researchers determine that the time is right, they send an unmanned combine harvester to mow the wheat and shoot it into the back of a grain tractor that drives itself alongside the combine.

Hands Free Hectare is just one of many examples of model farm projects that aim to demonstrate the promise of automated, autonomous agriculture on a large scale. In the US, the Microsoft-backed Grand Farm outside of Fargo, North Dakota is building a farm complex to experiment with robotic planting and irrigation; the University of California, Davis' Smart Farm initiative is building "an Agricultural Innovation Hub to develop smart machines with industry collaborators." Fully autonomous agricultural machinery is not yet commonly used, or even functional outside of very constrained environments. But the GPS-controlled machinery and drones that make that vision possible to imagine and fund today have already been widely adopted by agricultural operations all over the world.

The adoption of technologies to automate and generate data about farming operations is known in the industry and in academia as "precision agriculture." The cheerleaders of this approach frame it as the optimization we need—"each plant gets just the right amount of water and fertiliser for maximum yield," as the *Financial Times* put it in 2017—to feed the world's billions of mouths and conserve resources in the face of climate change. Precision agriculture is heralded by development agencies and funders as a solution to food insecurity. Under the banner of "feeding the world" the digitization and datafication of agriculture have gathered so much momentum that their continued development and adoption seems inevitable.

But the most promising takes on precision agriculture rarely mention the numerous threats that accompany it. All of that efficiency often requires not just sensors, but coordination among the sensors, routing of the data that those sensors collect back to a central data store, and ongoing monitoring to ensure that batteries haven't died and connections haven't dropped along the way. The proliferation in agriculture of systems that are supposed to run themselves has created a massive surface area of weak links, delicately strung together.

Threats to food supplies have always come in many shapes and forms: famine, war, and sabotage have wreaked havoc on societies throughout human history. Now, those of us whose access to food is mediated by digitized agricultural production systems face a double threat: the rapidly expanding mesh of insecure devices that control how our food is planted, watered, monitored, harvested, and transported; and, at the same time, the shifting balance of power away from farmers, towards the governments, corporations, and investors that wield massive amounts of capital.

> *"The proliferation in agriculture of systems that are supposed to run themselves has created a massive surface area of weak links, delicately strung together."*

Making Hay

In its purest form, the goal of precision agriculture is to guide or control every part of the food production system: what type of seed to use, how much fertilizer to apply, when to plant and harvest. The approach has its origins in the Geographic Information Systems (GIS) that was introduced on farms in the 1960s and 1970s to plot the locations of soil samples on a map, and in GPS receivers and mobile tools to monitor yields and soil composition, which were introduced in the 1990s. More recently, the rise of precision agriculture has been enabled by the explosion in internet connectivity that the world has experienced over the past ten years.

Farmers historically have been early adopters of sophisticated agricultural technologies, driven as much by incentive as by need. Governments often provide subsidies to digitize agricultural equipment and methods. Today, almost every piece of agricultural machinery sold by international corporations like Kinze, Deere, and Case IH has navigation systems and a variety of sensors preinstalled on it by the equipment manufacturers. Farmers can supplement these preinstalled systems with additional receivers on equipment like sprayers and combines, as well as ground sensors and drones that collect data points on moisture levels, mineral levels, and crop growth. Even smallholders—those farming on less than ten hectares, or 0.04 square miles—are using these technologies in places like Nigeria, Colombia, and Indonesia.

> *"Security has not been a hot topic at all recently as we are still getting the fundamental systems working. Mostly we just use Wi-Fi."*

McKinsey & Company considers agriculture to be a "massive opportunity" and an area "ripe for disruption." Goldman Sachs expects the market to soon be worth $240 billion. Despite questionable returns, agriculture has become a darling of VC; according to the VC fund AgFunder, agricultural technology startups secured nearly $17 billion in funding in 2018, up 43 percent from 2017.

Academic institutions, too, have turned their prowess towards fostering the development of new agricultural technology through startup accelerators and enhancing technologies of established corporations. Indeed, according to food systems expert Kevin Walker, academics in the US are becoming subcontractors of large multinational corporations, driven by the need to find funding for research that has "immediate and marketable benefits."

The US government and military are also invested in the research and development of precision agriculture tools. The US Navy has invested in agricultural robotic swarm technology. Many of the startups developing drones for commercial agriculture are run by military veterans or former defense contractors. Even economic development sector initiatives rely upon defense contractors.

These types of funding streams yield specific results. The demand by VCs and universities for exponential returns in a short time frame means that security is, as with many non-agricultural venture-backed software startups, an afterthought. Hardware is adapted from other sectors and software outsourced, with no plan for service, repair, or maintenance, leaving pieces of equipment perpetually vulnerable to exploitation when it reaches end-of-life. And, even if they wanted to, private companies trying to translate government research and development into consumer agricultural products don't have the billion-dollar budgets to prioritize security in the same way that DARPA might.

Breach Party

Many of the same security vulnerabilities that plague battery-powered, cloud-connected devices in general affect farmers adopting precision agriculture. In an interview with precision agriculture expert Marc Window, a professor associated with Hands Free Hectare said of the project's approach to security,

"As with most of our ag robot developments we use technology that has been developed outside the ag area and migrate it over as and when needed. Security has not been a hot topic at all recently as we are still getting the fundamental systems working. Mostly we just use Wi-Fi." This approach to (not) securing connected agricultural machinery is the norm, and it means that, in the same way that corporations can remotely brick a piece of machinery, malicious actors could hypothetically do the same, bringing a fleet of tractors down at once. One can imagine a scenario in which a piece of code is deployed to disrupt the harvest of entire nations. Or a scenario in which chicken farmers who use web-based software to remotely control the temperature of their hatcheries find their cooling systems manipulated, killing their animals.

These are not purely hypothetical. China's environmental efforts are being thwarted by companies doctoring surveillance camera footage and remotely altering or deleting undesirable information in automatic monitoring systems in order to appear more environmentally friendly than they are. A smart irrigation system in Israel has reportedly been hacked by the Syrian Electronic Army, a hacker group. Smart irrigation systems in general can be manipulated to empty entire water reservoirs or apply the wrong amounts of fertilizer and chemicals.

Drones are being heralded as revolutionary in agriculture, but they have many known vulnerabilities since their security is essentially unregulated. They can be hijacked, remotely tampered with to return false data, or piloted to infiltrate remote Wi-Fi networks. The US Department of the Interior also sees espionage risks in the fact that its own drone fleets are manufactured by a Chinese company.

The tools of precision agriculture generate valuable datasets, and the value of those datasets corresponds to their size. To give a sense: the largest dairy farm in the world occupies more than 22 million acres and the largest field operation sits on more than 500,000 acres. Data from these operations can move markets. The US Department of Homeland Security has reported that at least one company has been approached by commodity brokers with an offer to buy the company's data. But insider trading threats aside, the integrity of large operations' data can be undermined by weaknesses in any number of factors that make precision agriculture work—IoT devices, software, or physical hardware. Although such market sabotage doesn't appear to have happened yet, agricultural data breaches have: the security industry's annual data breach compilation identified eleven data breaches among agricultural companies in 2019.

> *"One can imagine a scenario in which a piece of code is deployed to disrupt the harvest of entire nations."*

Even as US government agencies propel the adoption of precision agriculture, they recognize the new security threats that it poses. In 2016, the FBI released a joint memo with the USDA warning farmers that they were increasingly at risk of having their data held for ransom or of bulk data theft. In 2017, the US Navy conducted war games designed to train servicemen to protect key sectors, including food and agriculture, against malicious state and extremist actors. Soldiers were trained to mitigate attacks like remote tampering of temperature readings at a vegetable canning facility and ransomware targeting an agricultural company's financial data.

Roll Your Own Milk Sensor Backup

But war games won't help farmers whom precision agriculture has made increasingly dependent on complex systems beyond their control: on the software that runs farming equipment, which can require paid subscriptions and constant updates; on the creditors who own the equipment, often worth hundreds of thousands of dollars, and to whom farmers must make regular payments; on manufacturers who have access to the data the equipment generates; and on the guarded knowledge of licensed repair technicians who are the only ones able to diagnose problems with proprietary systems.

Take the relationship between farmers and John Deere, the largest manufacturer of agricultural machinery in the world. For years, John Deere has put proprietary software and hardware in its tractors, but now those tractors are increasingly connected to John Deere's servers as well. According to an executive with the company's innovation arm, "Our large equipment now has 4G LTE modems, with Wi-Fi and Bluetooth, and that does two-way communication so it collects data off your farm and sends it to the cloud." As a result, the company can see all the data collected by the machinery they manufactured and can also remotely lock down, or "brick," equipment if they suspect that farmers are trying to repair the equipment in unsanctioned ways. Farmers have responded with the Right to Repair movement on the basis that they shouldn't have to fear having their own tractors remotely bricked for any reason, much less for making repairs.

Or consider Emiel Stam, a Dutch dairy farmer who used AgroVision robotics to milk his cows. He claimed in Dutch court that one of the robot's sensors malfunctioned, infecting his cows and forcing him to cull his cattle. To make his case, Stam needed a historical record of the offending sensor's readings—in this case, the milk flow sensor.

Stam had backups of those sensor readings through the machinery's software interface, but they had been remotely wiped by the robotics company, which had access to the readings through a maintenance contract. Luckily, and very unusually, Stam had additional independent backups. He used his own data to correlate the milk sensor readings with the onset of his cows' illness, prove the robotics company's liability, and win damages for his losses. But he is the exception that proves the rule.

"War games won't help farmers whom precision agriculture has made increasingly dependent on complex systems beyond their control."

Beyond 2FA

Now that food systems globally are increasingly vulnerable to digital manipulation, farmers need to protect themselves. The FBI's 2016 joint memo with the USDA encourages both small and large-scale agriculture operations to implement standard digital security measures: using secure passwords, setting up two-factor authentication, accessing networks via VPNs, and having company-specific email accounts for employees.

But perhaps adding more technology to the technology problem is not the solution. One issue that extends well beyond agriculture is that there is no clear line of liability for the security of equipment and devices. To help demarcate where the responsibility lies, venture funds, development agencies, and government procurers could implement something like the US Department of Justice's new policy on drone use. It requires partners and grantees to conduct a mandatory cybersecurity risk assessment and honor a data retention policy. Imposing such requirements on small organizations would be a financial burden, but perhaps such a cost is worth paying to protect farmers and food supplies.

While these steps are necessary, they are not sufficient. We have a globally connected food system, and ameliorating harm requires systems thinking. Hacks are inevitable when we use connected technologies; the more we become reliant upon them to bring in our harvests, the more we can be assured that these systems will be exploited.

Rian Wanstreet is a PhD candidate at the University of Washington and an Open Science Fellow with Mozilla, researching the ramifications of new technologies in the agricultural sector.

Party at My House

Darius Kazemi on Human-Scale Social Media

Over the last few years, there has been a slow drift away from major social networking sites like Facebook, Twitter, and Instagram. Some people left in order to to declutter their digital lives, while others left because the platforms no longer provided the thing that they came there seeking: community. But if you're seeking community, where do you go?

It turns out: someplace that looks very similar. But unlike those big social media sites, where the rules, moderation, servers, and your data are controlled by Facebook or Twitter, this place is controlled by you and your friends. There are no ads, the code is open-source, and services are built on open standards. As a concept, it's called the decentralized web: run your own piece of social media software, but let it communicate with others and, most importantly, make and enforce your own rules, both technical and social.

Darius Kazemi is an internet artist and software developer who thinks the internet can be a better home for humans. He recently wrote a best practices guide called "Run your own social: How to run a small social network site for your friends" (https://runyourown.social) to talk about why and how you might build a community on decentralized social media. His purpose was to provide another way forward—and explain it to anyone, whether you're able to stand up your own server or not.

"Run your own social" focuses on Mastodon, the most popular open-source software for decentralized social media. Through open protocols like ActivityPub, Mastodon servers are able to communicate with the larger "fediverse," the larger ecosystem of interconnected—or federated—servers. From a user's perspective, Mastodon looks and acts a lot like Twitter, but it's able to connect to a wide range of fediverse services similar to YouTube, Instagram, Blogspot, and more.

Darius's guide is based on his experience running his own small social network site for his friends known as Friend Camp, where a few *Logic* editors are members. In the chat that follows, conducted in January 2020, we talked to Darius about Friend Camp, the fediverse, and, briefly, our attempt to create an instance of our own.

How did you get interested in running your own social network site for your friends?

At some point Twitter became a hostile place for me to put my bots, and I became interested in the tech platforms that are not owned by anyone in particular. I got started with my own Mastodon instance just for myself for a while, but it got lonely. At the same time, I had a whole network of friends on Twitter who all had locked accounts and would primarily talk to each other there, because we all trusted each other. So I figured: I have tech skills and social skills. People trust me. Maybe I should try running an instance for my friends. That's what Friend Camp started out as—an experiment in trust, as much as anything else.

Then I started making modifications to the Mastodon software that Friend Camp was running on, which turned into the offshoot that I call Hometown.

When you say Twitter became a hostile place for you to share your work, what do you mean by that?

First, it became technically difficult to maintain my bot projects. Twitter changed their API around a lot, and were bad about communicating the changes, and which features would stay and which would go. That meant that whenever I noticed a change, I suddenly needed to update all eighty of my Twitter bots to get them working again.

I also got a bad taste in my mouth from Twitter supporting kinds of speech that I don't particularly agree with, and I began to feel like putting my toys and art there was just adding value to the network. For a while I felt that the tradeoff was that the art I published there still enriched my life and the lives of my friends. At some point, though, I started to feel like the balance got out of whack, and I could do more good by putting my work on alternate networks and giving people a reason to get off Twitter.

I joined a Mastodon instance a couple of years ago, and then fell off using it. I'd remember it every once in a while and think it was cool, but still not end up using it for long. Then I joined Friend Camp, and I loved the community, but still wasn't an active contributor.

The guide you published about running a small social media network for your friends, "Run your own social," got me interested in the idea again. What led to you write the guide?

That project was the culmination of my Mozilla Fellowship. I didn't know exactly what I wanted to do for the fellowship going into it, but I did know that I wanted to work on decentralized social media.

Through that, I got to meet a bunch of people working on all sorts of different protocols who were active users of different kinds of networks. The early adopters were people who believed in certain principles around free information. Those ideas are interesting to a very particular kind of person, but they're not that compelling to everyone else. And the more I learned about and participated in decentralized social networks, the more I realized that what was interesting to me about them—the things that made them cool and different—no one was talking about.

> ***"To me, [internet health] is a really moralizing stance. It's a metaphor like eat your vegetables, lose some weight, here's what you have to do to live a morally better life through technology."***

So I wanted to write a guide that talked about why someone who isn't primarily a nerd should be interested in decentralized information flow. What does it bring to your life? And not only that, but what does it bring to your life that Twitter and Facebook don't?

It was also a reaction to what I saw in tech liberation spaces, including Mozilla, where the guiding metaphor is "internet health." To me, that's a really moralizing stance. It's a metaphor like: eat your vegetables, lose some weight, here's what you have to do to live a morally better life through technology.

That works for some people, obviously. You know, people care about their health and there is a big industry around health advice, so presumably there are people who care about that sort of thing in a tech space as well. But I wanted to imagine a way of having fun on the decentralized web and bringing value to your life in ways that you can't on the centralized web.

What finally pushed me to create my own Mastodon instance was feeling that it was possible to build a special community by incorporating certain moderation features that weren't available on the regular Mastodon project. The most important feature for me was the local-only posting, which allows users to create posts that can only be seen by members of that instance.

What are the technical features you've built into Hometown that you feel are needed to create the kind of community that you want to support on the web?

I plan to add more in the future, but Hometown currently has three features that are different from Mastodon. Local-only posting is the main one, and I didn't even write that feature. It was originally written for glitch-soc, a fork of Mastodon with more experimental features, and a Brazilian guy named Ronato "Lond" Cerqueira submitted it as a pull request to the main Mastodon project—which has never been merged. I basically just took the feature with his blessing. It's not technically hard—all it is is an extra true / false column in the database, and then some extra logic that takes that into account. I haven't counted, but I would guess it's around one hundred lines of code, on top of the massive Mastodon code base. It's a very small change to Mastodon itself, but it strikes at one of the core things about Mastodon, which is this concept of federation and relaying messages to other places.

There are lots of Mastodon instances out there that are general interest, where people join who don't necessarily have anything to do with each other. It's kind of like your choice of email server—you're not in a community with other Gmail users because you have a Gmail account, you're not in a community with other Hotmail users because you have a Hotmail account. Except that on Mastodon, there are features that seem to *imply* that you are in a community, like the instance-specific timeline, also known as the local timeline. But the local timeline doesn't make that much sense if there's nothing that you have in common with anyone else.

> *"My promise doesn't mean anything to a random person off the street. But it does mean something to you, because you're on Friend Camp."*

Then there are instances that are built around particular fandoms, and that can sometimes work. But I'm more interested in people who already have preexisting social connections to one another. For a group of preexisting friends, it made sense for me to provide an extra layer of security, where only the other people in the group can see my messages.

I like to go back to email as an example over and over, because email is federated. It has been forever. I can have a Gmail account, and you can have a Hotmail account, and I can send you an email because our services speak the same language, even though our accounts are on different servers and may be using entirely different technology. Federated social media is basically the same idea: two servers that speak the same language, and enable people to talk to each other.

Email also has similar questions of trust as federated social media. If I send an email to, you know, becky@gmail.com, it's going to the Gmail servers and Google is going to see the message and they're going to do what they're gonna do with it. If I trust Google, that's okay by me. But if I go to my auto mechanic and I send an email to contact@jimsmechanic.com, I don't know if that email is run by Gmail behind the scenes, or if it's custom technology or what. For all I know, every email that goes to that mechanic's domain is also posted to a subreddit somewhere that people use to target other people for harassment. I don't know!

That's how posts on the fediverse work too. Friend Camp is connected to something like 6,000 other servers. I haven't audited these 6,000 servers to know if they're safe, nor could I. If my messages go out to any one of those servers, there could be someone opening up private messages and looking at them.

So the local-only feature is about a topology of trust. I know that everybody on Friend Camp trusts me with their data, so if we can have a local-only layer, that means there's only one computer that stores these messages and that's the one that's owned by me, a person that you trust. So a message that you send to someone else on Friend Camp goes from your phone to my computer that's owned by me, who you trust, and then it goes to your friend's phone, and you trust your friend too. That provides an extra layer of security. So for example I might tell people: if you're going to send nudes, send them local-only, and I promise I won't look at them. My promise doesn't mean anything to a random person off the street. But it does mean something to you, because you're on Friend Camp and we have a trusted relationship.

If you made these communities available to a general audience, would the trust part still work? In other words, do you think that these notions of trust and security scale?

I've spent a lot of time thinking about that. Not everybody knows a Darius—not everybody has a friend who's technical and wants to set up one of these networks, and there's currently only one hosting company you could go to who would set this kind of thing up for you.

But this technology could easily piggyback on existing trust networks and infrastructure networks. As a thought experiment, what if every public library system in the US had a federated social media server, and anybody with a library card could have an account? That's not as personal as something like Friend Camp, where the administrator is your friend. But if your administrator is an employee of your local library, then there's a door you can go knock on. You can complain to your local city council member if something happens that you don't like. I can't go knock on Mark Zuckerberg's door and complain about something, but I can do that at the local library.

That would still be a very small-scale, local thing. I would love to see municipal infrastructure take on some of this, at least in a US context.

I'm still trying to figure out how this works at a larger scale. Let's say one billion people want to be on social media. If we're thinking about my ideal of small communities, that would mean that ideally there would be something like ten million servers of a hundred people out there. But how does that even work? How do you have ten million servers? And does that mean you only get to have a hundred people in your local-only posting club?

This is where I am playing around with concepts like "neighborhoods," where a bunch of servers mutually agree to trust each other. That would allow for three levels: you could have local-only posts that only go to your instance; you could have neighborhood posts where messages are only federated to other servers in the same neighborhood; and you could have your global public messages that go to everyone.

"What if every public library system in the US had a federated social media server, and anybody with a library card could have an account?"

This reminds me of the Community Memory project that was established in the 1970s. It was a system of public computer terminals around Berkeley and San Francisco, where anyone could go to whatever physical terminal was closest to them to see what other people had posted, check for messages, and leave messages for other people. That idea of public, decentralized infrastructure that's rooted in a particular place feels very similar.

Yeah. It's kind of similar to bulletin board systems (BBSes) back in the day, too. You wanted a local BBS in your area code because otherwise you'd have to pay long distance telephone fees. Honestly, if federated social media becomes as successful as BBSes and lasts for about as long as they did—maybe ten years of a heyday—that would be a huge success.

There is a book called *Cyberville* by Stacy Horn about the local BBS community she ran in New York City for something like ten years. The book was written in 1998, and it's really relevant to managing a fediverse instance.

Sometimes I try to think about how to get the whole world on these kinds of decentralized networks. But on the other hand, I question why the whole world needs to be on these kinds of networks. I can't think of a compelling reason—other than the selfish one of wanting to talk to everyone that I know. But you don't always get everything you want.

Your experiments are able to demonstrate a sense of trust online that isn't a binary, on/off thing, but where there are degrees of trust. That's common in interpersonal relationships—you may have an inner circle, then acquaintances, and maybe you have a public persona—but it's largely absent on the web.

Even though your projects are built on Mastodon, it seems like the maintainers of the Mastodon project have a different vision for this system of federation—which is not necessarily to build up networks of trust, but to instead to create an interconnected world. And despite Mastodon having a very different structure from Facebook or Twitter, it feels like all those services share that same motivation.

That's my primary critique of Mastodon: that it's still buying into the fact that everyone talking to everyone else is necessarily a good thing.

It's not just Mastodon—most fediverse developers out there seem to be really interested in the idea of connecting everybody in the world to everybody else in the world. I'm not. I'm trying to create a counter-narrative to that with "Run your own social" and Hometown. And more and more people are switching from Mastodon to Hometown, which is really cool, and a little scary for me. But my hope is that Mastodon will just accept the pull request for my Hometown features, at which point we could all move back to Mastodon.

> *"Most fediverse developers out there seem to be really interested in the idea of connecting everybody in the world to everybody else in the world. I'm not."*

Everything In Moderation

You've crafted a very particular code of conduct for Friend Camp, and you write in the runyourown.social guide that if someone can't disagree with your code of conduct, then it's probably not strong enough. How do you think about moderation?

The thing about keeping a server and a community small—and by small, I mean less than a hundred people—is that it is possible to individually know every single user and have a sense of where they're at, how they're doing, and really take care of everyone as needed. It also becomes possible in a community that small to make decisions with a pretty strong consensus, if not unanimous agreement, and it's easier to engage in difficult conversations.

Just this morning, a fairly major moderation request happened on Friend Camp. Someone made a request to defederate two particular servers, which is essentially a server-wide block on sending and receiving messages from those servers. So that would mean that if I had friends on server A and B, we would not be able to communicate with each other anymore over the social network—we wouldn't even be able to see each other's content.

> *"If your goal is to get people off of the centralized services and onto the network that you're on, the only way to do that is to provide better service."*

Typically, the easiest defederation requests are for servers that are full of spammers, or people who are making death threats—that's the cut and dried stuff. For the defederation request that came in today, which was reasonable but less obvious than the examples I just gave, I was able to go into the database and look up exactly how many people we follow on the servers in question. We had three active Friend Campers who followed an assortment of about a dozen different accounts between those two different servers. From there, I looked at each one of those dozen accounts to determine which ones were active accounts, and wrote a custom message to each of those three people on Friend Camp. I showed them the thread about defederating and explained that we hadn't made the decision yet, but laid out what the consequences would be in terms of not being able to talk to the specific people they followed. It's possible to have informed consent in a community of this size—I'm able to contact the people who are immediately affected and make sure that it's brought to their attention. That doesn't scale.

So that's the kind of thing that can't be offered by Twitter or Facebook, or even larger Mastodon instances like mastodon.social, which has something like 200,000 users. It's a "feature," in air quotes. It's enabled by the technical details of how federation works, but it relies heavily on the labor of a moderator and the engagement of users—so it's more of a social feature than a technical one. Ninety-nine percent of Mastodon and other federated servers don't have admins who do this sort of detailed moderation.

That version of content moderation, where you are stewarding the norms of a community, is so different than the Big Tech version, which feels like it is mainly about removing terrible videos of people doing terrible things. Is there any way to scale or automate your approach?

In a way, it's like asking if something as personal as sex work could scale. And, in a sense, live-streamers creating a one-to-many relationship is sex work at scale, but it's a very different thing than one-to-one.

I think that the one-to-one dynamic is what differentiates federated social media from platforms like Twitter. And if a federated community doesn't operate this way, then there's no reason for people to not just be on Twitter. If your goal is to get people off of the centralized services and onto the network that you're on, the only way to do that is to provide better service.

There is a narrative that is invoked with 4chan, 8chan, or Reddit, where if you have a community that creates its own

rules and is not subject to some centralized moderating force, you can actually end up radicalizing that community. Do you think that there's something to that?

I believe in radicalizing people. I don't think there's anything wrong with radicalization. I can disagree vehemently with the views of a given radical community. I've heard people say that there are no Nazis on Mastodon. I'm like, what are you talking about? There are so many fucking Nazis on Mastodon. But on Friend Camp, *I'm* here, and when I hear about the Nazi servers I defederate with them. Another example is we have lots of trans people inside Friend Camp and so we have a zero tolerance policy for trans-exclusionary radical feminism. We just don't accept that.

Is that the right policy for Facebook, where they're trying to serve the entire world? I don't know. It's not an obvious call if you want to serve every single person in the world. But if the fifty of us who are active Friend Campers decided that mentioning peanut butter and jelly sandwiches is completely against our values as a community, and we want to defederate with anyone who mentions them, we could do that. We have that ability.

So I'm not against radicalization. I am fine with having a radical queer community with ideas about social justice that seem really strange to outsiders. But I also think you can't allow for radical communities of one kind, at least technically speaking, and not also enable spaces for radical communities of a different kind. Nazis will also be able to run their own Nazi instances with Nazi rules. I think about it in terms of my home—there are people's houses in my neighborhood where I can see in the damn window that they have Confederate flags up, and I'm sure they have poker nights where they get together and talk about shit that I really don't like. But they're doing it in the privacy of their homes. I think you can assemble in a private space and talk about whatever. It's when it comes out and enters the public forum that consequences can start coming into play.

> *"You can't allow for radical communities of one kind, at least technically speaking, and not also enable spaces for radical communities of a different kind."*

Oh Hi, Mark

You take an hour to three hours to walk new users through Friend Camp, describing the features, and connecting them to others on the network that they may not know. That's something that you don't get on larger platforms. The process also seems akin to consciousness-raising or educating people about their own data. Could you talk about your approach to onboarding, and why you decided to do it that way?

To be clear, I spend an hour, sometimes two hours—never three, fortunately.

Ah, in your guide you said up to three. I was thinking, my God, that's so much time.

I was just putting out a worst-case scenario for people so they know what they're getting into. But, yeah, I basically go through every single technical feature, every button that you can press, every preference in your profile. And I am there in the future if they ever have any questions about how any of this stuff works.

You can have complicated features that are not super easy to use, as long as there's someone there who can help you through it. I'm that person for Friend Camp. Because I personally know every single person that joins Friend Camp, I also have some idea of what they're interested in, so I tailor it a little bit to each person. If I know someone's really privacy conscious, I'll spend more time going over privacy features with them in more granularity.

Facebook's privacy settings are actually pretty comprehensive, but it's this massive dashboard that's confusing to people who aren't technical and don't know what all the different words mean. On Friend Camp, we also have complicated dashboards—but you can also message the Mark Zuckerberg of Friend Camp and ask him to explain what all the buttons mean. That's as far as I'll compare myself to Mark Zuckerberg.

The second half of the orientation is really a social orientation. I like to think about running a federated social media server instance as being like the host of a party. If someone shows up to my party who's just moved to town, it's my job to take them around and introduce them to people and say a few words about them, and maybe take specific time to introduce them to individuals who I think that they should meet. Usually what that means is I go on the local timeline and I scroll down, and for everybody who's been actively posting for the last twenty-four hours, I describe in one sentence who every single person is, and go into more detail if I think there's a potential connection there.

This ties in with another reason why I wrote my guide. People talk about how it's hard to get people off Facebook and Twitter because of the network effect. People want to be where the people are, right? With the guide, my hope was—and it seems to be borne out, people are doing this—to bring *whole communities* over at a time rather than individuals. I think trying to recruit individuals away from a social network site is a fool's game, because you invite them over and maybe they know you, but they don't know anyone else. That's it. Whereas if you can get an entire church group or an activist group to move over to decentralized social media together, then you don't have that problem. Their community is already here. There's already dozens of people that they know.

> *"I like to think about running a federated social media server instance as being like a party host."*

Now, this level of care for a social media instance is a big commitment—between onboardings and support, or fixing things like Elasticsearch being down, I spend something like twenty or thirty hours a month on it. I try to make it clear to people: it takes up as much time as any major hobby. You do this instead of watching TV, or really anything else.

Just having started my own instance, there's a question of how we go from zero to a vibrant community. I want to build agency and context for folks when they come on to the instance, which I think requires doing the kind of onboarding that you're doing. But I don't have the time to onboard a whole bunch of people right now.

Right. In that scenario, I'd consider just doing a big group call. Just have two or three group teleconferences—dial in and get the onboarding. Getting over that initial hump is really, really important. And it requires a ton of work. It's a lot, a lot, a lot of work. But I think it's worth it.

Break Up with Your Platform

I want to make sure we get to talking about the role of open standards. I'm relatively new to decentralized protocols like ActivityPub, Scuttlebutt, and those sorts of things. How do you see those different protocols supporting potential futures that may exist on the social web?

Protocols are just agreements: social contracts between people saying we're going to speak this language to each other, and we abide by these rules. Some protocols have so much value that everyone uses them. For example, everyone uses TCP/IP, because otherwise you don't get access to the infrastructure of the internet—the actual wires that move data around from place to place. That's a very compelling reason to use a protocol.

> *"Different protocols map onto different ideas about what it is to be human and communicate."*

When it comes to new decentralized protocols like ActivityPub and Scuttlebutt and Dat, there isn't yet a compelling reason for everyone to use them. But what's interesting is how these protocols encode different ideologies. My favorite example of a protocol encoding an ideology is Morse code. Morse code is an early technical protocol for communicating over the telegraph wires. And one of the ideologies encoded in it is an English-centric system, because there is no Morse code for an "A" with an accent on it. Morse code only covers the English alphabet, and that's it.

With the decentralized protocols, some of them encode ideologies around privacy. I would say Scuttlebutt is more libertarian than ActivityPub, because it's about every individual being their own server. And ActivityPub was interesting to me because it maps more towards my ideologies, which is that we all need to band together and share resources, but that it's also okay to have some specialization.

Mostly, though, open standards are a way for me to easily take advantage of stuff that's already out there. It's a way for me as an individual to amplify my time and effort as a developer. I'm about to release an event management project for the fediverse. And one of the cool things about it is that while I built it to be compatible with Mastodon, because I was doing that through the open standard it already kind of works in a bunch of other ActivityPub-based services. It didn't require that much extra effort for me to tweak it to make it work for systems other than Mastodon, and it actually works better in some of these other services than the ones I designed it for.

The different protocols map onto different ideas about what it is to be human and communicate, really. And they make different assumptions about users. It's like, well, *of course* you want to be your own server. But maybe you don't. Maybe it's more ecologically efficient for you *not* to be your own server, and for people to share them. When these protocols state their assumptions, they tend to state them as universal truths. It's kind of like the way product marketing works: don't you *hate it* when you crack an egg and it explodes in your face? Buy my new egg cracker! It's kind of like that.

Jack Dorsey announced in late 2019 that Twitter is funding five people to develop an open standard to decentralize Twitter. What do you think?

For a huge company like Twitter, it's easy to fund five people to work on a project for a few years. If you do the basic math, you could fund five people for five years with less than ten million dollars. That is obviously a shitload of money that could change a lot of people's lives, but in the grand scheme of Twitter, it's not very much.

So I think it's just a hedge. It's a way for them to tell investors that they're not ignoring the space; you can't say they're not paying attention to it. But honestly, it also strikes me as the kind of thing where, at a big tech company, internal politics can drive the creation of new initiatives. Maybe there's a handful of people who you want to keep on staff and who are at risk of leaving, so you give them a cool project to work on and then they work on that cool project for five years. There can be many reasons for a big company to do this sort of thing, but I don't see it as a serious initiative. And even if it is serious, like all initiatives it has a low chance of producing anything fruitful. I'm not going to worry about it until I see them start making moves that actually make a difference.

There have been moves in the US and Europe towards regulating these large platforms. Many folks have called for them to be broken up. What do you think should be done with them?

I would love to see them all broken up. I think that would be a start. I'm not against the existence of giant centralized platforms, as long as it's a real choice as to whether you get to be on one or not. Right now, it really feels like it's not. You don't have a choice to be on Facebook or not—not really. I don't have a Facebook account. I haven't had one for a year. But there are things that I miss about it, and I'm channeling that and using it to build an alternative.

> *"You don't have a choice to be on Facebook or not—not really. I don't have a Facebook account... but there are things that I miss about it, and I'm channeling that and using it to build an alternative."*

I haven't had a Facebook account for a bit longer than that. The things that I miss are mainly interactions with friends and family who aren't as interested in experimenting with new technologies, which right now feels like a barrier to entry to using Mastodon. What would it take to have Mastodon or these decentralized networks become more mainstream?

I like the small-community approach. I think it would be possible to bring on people who aren't technically advanced, as long as you have an individual coordinating the whole thing.

And one of the beautiful things about federation is that you can have entirely different services talking to each other. So you could have Friend Camp talking to an instance that does

Facebook events-type stuff, and you can also be talking to PeerTube, which is a YouTube-type service. There's nothing preventing someone from making a federated app that, on the front end, looks exactly like a text messaging interface that you could use to talk to your grandparents, because most grandparents can send a text message. Now, is anyone doing that? No. My hope is that more people will.

A Brief History of the Gig

by Veena Dubal

Where does the gig economy come from?

In early 2012, San Francisco taxi drivers began to raise the alarm at organizing meetings and city hearings about "bandit tech cabs" pilfering their fares. "I'll sit at a hotel line, and I see one of these guys in their own car come up, hailed by some guy's app, and they'll turn down my fare," Dave, who had been driving a taxi for fourteen years, said at a meeting that April. "They steal it. It's insulting." Other cabbies said they were seeing the same thing, and that their income was suffering as a result. "I think I made 20 percent less last week than normal," one driver lamented. He wanted city officials to know; surely, they would put an end to it.

These taxi workers were experiencing the very earliest stage of a global re-organization of private and public transportation, fueled by billions of dollars of financing from venture capitalists. Under the shadow of the Great Recession, in a period of high unemployment and slow job growth, Uber, Lyft, and their erstwhile competitor Sidecar used this technocapital to begin offering almost anyone with access to a car a way to make money by driving people around San Francisco. The companies aggressively marketed themselves as disruptors of the transportation industry; consumers and commentators, seduced by on-demand, technology-fueled mobility and the prosocial promise of what the companies called "collaborative consumption," enthusiastically adopted their narrative.

> *"The ride-hailing industry is facing a wave of militant self-organizing and claims to employment status by drivers."*

Cabbies, however, saw Uber and Lyft as well-financed corporate continuations of the taxi companies that had long subjugated them. "This isn't about technology," Mark, a long-time taxi worker and advocate told me in 2013. He explained that, for the previous few years, San Francisco taxi drivers had already been using an app called Cabulous, which essentially did what Uber and Lyft were doing. "They claim they're innovative and new, but we already have this technology," he went on. What was different, Mark described, was that Uber and Lyft had "a new exploitative business model," though it was just "one step removed from the leasing model that the taxi companies have been using for years."

Since 2012, much of the positive discourse around Uber and Lyft has continued to regurgitate the notion that these are companies built on technological innovations that brought new forms of transportation to people and places who needed them.

Meanwhile, critiques of these companies, and of the gig economy as a whole, have typically seen Uber and Lyft as breaking sharply from earlier modes of employment to create new forms of precarity for workers. In both cases, the public discussion tends to see these companies as creating major discontinuities, whether of technology or of labor models. What Mark pointed out, however, is that Uber and Lyft are in many ways not as different as we tend to think from the taxi companies that prevailed until 2012.

In 2020, almost a decade after the advent of Uber and Lyft, we seem to be at another turning point. The ride-hailing industry is facing a wave of militant self-organizing and claims to employment status by drivers. So far, the most significant mobilization has been the fight over AB5, a California assembly bill that was signed into law in September 2019, and which makes it much clearer that drivers should be treated as employees of Uber and Lyft. The companies have fought this reclassification in myriad ways, and some drivers fear that it may cause them to lose their flexibility. But those who have welcomed the passage of AB5 hope it will deliver them many of the benefits—from healthcare to a guaranteed minimum wage—that Uber and Lyft have so far denied them. On all sides of the issue, no one doubts that we are at a critical juncture in the history of labor and urban transportation.

But in order to sort through the arguments surrounding AB5 and grasp the significance of this moment, we must do something that the discourse around ride-hailing has failed to do: situate ourselves historically, tracing both the continuities and the discontinuities that the cabbie Mark pointed to. Our present moment is largely the product of two neoliberal shifts in the taxicab industry—and, in a certain sense, in US society as a whole—that occurred in the late 1970s and the 2010s.

Understanding the reasons for these shifts can help us get beyond the easy assumptions made on different sides of the debate: that employee status is an unalloyed good or ill, that innovation made the rise of Uber and Lyft inevitable, or that the issues raised by the sector are matters of technology rather than politics.

Few people understand those reasons better than the drivers themselves—though, like other workers, they rarely have their voices centered in public discourse. By listening to drivers' accounts of how their industry operates and has changed, we can come to understand how and why, despite some fears and ambivalence, they are using employee status to create a much-needed friction in the wheels of technocapital.

> *"The independent contractor model that underlies today's gig economy first developed in the taxicab industry in the late 1970s."*

Leasing as Liberation

Typically, mainstream observers see two defining features of the business model of Uber and Lyft. The first is that the people who drive "on" their platforms are treated not as employees of the companies, but rather as independent contractors using those platforms to run their own personal taxicab businesses. In the US, direct employment increases corporate costs by roughly one-third, so classifying workers as independent contractors significantly increases profitability. (This is essentially the labor arrangement underlying the entire gig economy, from people delivering food and groceries to those performing rote tasks for Amazon's Mechanical Turk.) The second defining feature is the technology—the apps and the various pricing and dispatching algorithms behind them—that the corporations use to exert enough control over drivers in order to provide a more or less on-demand service. In both cases—the non-employment model and the technologies—Uber and Lyft share more in common with the previous generation of taxicab companies than many people understand.

The independent contractor model that underlies today's gig economy first developed in the taxicab industry in the late 1970s, as the United States shifted towards a neoliberal conception of society in which almost everything was to be subjected to the forces of competition. Workers and households were reimagined as entrepreneurial concerns, sole proprietorships that should fend for themselves in the tumult of the great American marketplace. Embodying this logic, the taxicab industry was one of the first among US businesses to slough off the costs and liabilities—minimum wages, healthcare benefits, disability insurance, among other things—associated with direct employment.

Taking advantage of rank-and-file discontent with traditional unions, as well as an existing carve-out for independent contractors in labor laws, cab companies all over the US reorganized their business models in this period, claiming to "free" their drivers from the imagined restraints of employment. In San Francisco, taxi companies approached drivers in 1976 and asked if they would like to "lease" their taxis on a shift-by-shift basis, so that the companies could rid themselves of the expenses and risks associated with employing workers, including unemployment insurance, workers' compensation, and a guaranteed commission that taxi drivers had previously been paid.

Although the Chauffeurs' Union, which had represented drivers since before the New Deal, discouraged workers from accepting these contract terms, many drivers were interested in the leasing model, which emerged within a tightly-controlled regulatory regime won by the union. In the early twentieth century, the union had successfully fought for municipal regulation of fares and restrictions on the number of taxis operating in the city. This helped to more or less ensure that drivers earned a living wage, even absent formal income guarantees.

Cabbies who chose to lease rather than work as employees continued to benefit from these laws. Many believed they could use their knowledge of the city and business acumen to earn more than they had as employees. Many hoped to be liberated from the companies' direction and control. Some had also grown distrustful of the Chauffers' Union, which took a politically conservative and dues-driven approach to the workforce. "The union was even worse" than the taxi companies, recalled Markos, a driver who had emigrated from Africa and signed a lease contract in the late 1970s, when I spoke to him in 2010. "They take your money and they say, 'Thank you and what can we do for you?'—and they didn't do nothing."

Markos and other drivers had also been frustrated by the union's positions on any number of political issues. Those who had embraced the civil rights movement or protested the Vietnam War had been ignored. Drivers who protested US interventions abroad by refusing to say the Pledge of Allegiance, which was recited at the beginning of union meetings, were ejected from the meetings. Dissident workers with long hair, whose left politics were in opposition to the union's, were even punished: ostensibly for their appearance, but more likely, for their political positions and agitation for a more radical union.

Though not all drivers initially embraced this shift, the decision was eventually made for them. When Yellow Cab, the biggest taxi company in San Francisco, went bankrupt and re-opened in the late 1970s, the company utilized a leasing model to contract with a workforce who, as independent contractors, did not have the right to union representation. Drivers who hoped to stay employees, particularly those with families, were alarmed. For the first time since the early twentieth century, taxi drivers in San Francisco were operating without a union contract and the safety nets associated with employment. Employment was "the only course for the working man if you want to have any kind of security, in the taxi industry particularly," Bill Williams, a thirty-nine-year-old driver with two young children told a local reporter at the time. "You can't afford the benefits individually… Your income varies, so you have to have a group plan, which is the union."

In response to this heightened insecurity, many cabbies organized and became increasingly engaged in city politics. For the next three decades, working without a union contract, taxi workers and activists formed informal worker organizations to lobby local politicians to maintain a regulatory environment that enabled a stable income. Drivers began regularly attending municipal meetings, and won a number of important victories for the taxi workforce, including getting a cap on how much the taxi companies could charge them to lease a cab. When workers' advocacy attempts were rebuffed—as in 2008, when drivers lobbied the city to institute a centralized dispatch system to improve service—they persisted. Many independently adopted the Cabulous ride-hailing app, which was strikingly similar to Uber and Lyft, so that riders could have access to the entire fleet (and not just the taxis at one company) when calling a cab.

Although drivers lacked any legal mechanism to collectively force companies to the bargaining table, workers continued to exert control over their income by opposing the issuance of new taxi medallions (thereby restricting competition) and lobbying for occasional fare increases. While some drivers, particularly the most vocal activists, wanted to fight for renewed employee status, the issue was contentious among driver groups. So, from the 1990s to 2010s, they avoided it.

As Ruach, a taxi worker and activist, complained to me about drivers' mobilization in this period: "We have to negotiate every fucking thing in our universe in public with the politicians and in the ballot box because we can't have a union because we can't have a contract because we are so-called 'independent contractors.' "

Remote Control

When Uber and Lyft began to eat into their income in the spring of 2012, San Francisco taxi drivers were relatively certain that they could rein in the "bandit tech cabs." Uber and Lyft were unequivocally violating both state and city laws. So, through organizing and advocacy, cabbies thought they could force the city government to enforce the laws and protect their interests, as they had done in previous eras.

And without even a fraction of the resources of technocapital, collectives of cab drivers put up an impressive fight. Between 2012 and 2014, in addition to organizing a number of demonstrations to protest the bandit tech cab companies, and writing and submitting sophisticated legal memos to regulators, taxi drivers also created a crowdsourced database of license-plate images of personal cars engaged in illegal ride-hailing. "We have tracked thousands of these cars," Barry, a key organizer in this endeavor, explained to me at the time. “We’ll share this with the SFMTA”—the local taxi regulator— “and insurance companies, because you know these guys are operating in violation of their personal insurance contracts.”

> *“‘We have to negotiate every fucking thing in our universe in public with the politicians and in the ballot box because we can’t have a union because we can’t have a contract because we are so-called ‘independent contractors.’”*

But these worker efforts were no match for the neoliberal political evolutions emerging in the US at the time. In the throes of the Great Recession, the individualizing, entrepreneurial logic that had emerged in the late 1970s intensified. For example, while the federal minimum wage remained unchanged, the Obama administration poured money into small business loans, with the hope that individuals taking on personal risk would reignite the economy. Rather than finding well-paying jobs, US workers were implicitly told to create them. Following this reasoning, in the taxi industry, state officials and politicians embraced the narrative that Uber and Lyft, in the name of micro-entrepreneurship, “freed” workers—not just from employment, but also from the taxi industry regulations (namely, vehicle caps and fare regulations) that had long kept their incomes stable.

But Uber and Lyft were taxi companies, just on venture capital steroids. The startups quickly created a loyal rider base by offering investor-subsidized prices and deceptively marketing themselves as a safer and more community-oriented service than regulated taxis. Though the companies were operating their ride-hailing services illegally, city regulators refused to intervene, citing the promise of technological innovation and a much-needed supply of jobs during the Great Recession's slow recovery.

> *"Wanting to avoid treating drivers as direct employees, Uber and Lyft nonetheless had to find ways to subtly exert control over them."*

Although less widely acknowledged, the regulatory pressure that Uber and Lyft exert—and the laws they flagrantly flaunt—is also a defining feature of their business model. In the early 2010s, by pouring money into California political campaigns (particularly that of Ed Lee, then the mayor of San Francisco) and hiring nearly every lobbyist in the state—including at least one who was later charged with illegal lobbying—the companies weaseled their way out of century-old regulations on fares, vehicle caps, and licenses in just a few months. They convinced state regulators to allow them to operate and then to carve them out of existing regulations by creating a new regulatory category, Transportation Network Companies (TNC).

Scandalously, the first TNC rules, adopted in September 2013, closely resembled those written by the companies themselves.

“They got exactly what they wanted,” Brad, a taxi driver said to me about the companies on the day the TNC regulations were released. “They wrote the laws themselves.” Indeed, the state regulations, which preempted local municipal regulations, seemed to inscribe the companies’ preferred business model into law. For example, even consumer safety regulations tied to professional licenses and state-monitored vehicles were conspicuously absent; instead, these safety oversights were outsourced to the companies themselves. Most critically for drivers, the new regulations lacked both any restraints on driver competition—in the form of vehicle caps—and state control over fares. The regulations were also explicitly silent on all labor issues, including work flexibility, a wage floor, the amount of time drivers could spend working per day, and most centrally, employment status.

Wanting to avoid treating drivers as direct employees, Uber and Lyft nonetheless had to find ways to subtly exert control over them. Rather than just tell drivers what to do and how to do it, which would trigger employment status, Uber and Lyft use “psychological inducements” derived from social science and deployed remotely through algorithms to influence when, where, and how long their drivers work. Instead of directly employing the millions of drivers whose work produces the company’s primary value, Uber hired hundreds of social and data scientists to shape driver behavior. (For example, using findings from social psychology and video game technologies, Uber notoriously tricks drivers into working at undesirable hours and locations.)

The success of Uber and Lyft in restructuring the taxi market and securing a legal environment that suited their interests emboldened venture capital to pour more money into the tech cab companies and to other on-demand labor platform start-ups, such as DoorDash and UberEats. As the technocapitalists celebrated these victories, a seemingly unlimited supply of drivers and declining commissions quickly devastated driver incomes across the entire taxi and ride-hailing economy. Almost overnight, full-time drivers made roughly 65 to 80 percent of what taxi drivers had made in 2012.

> *"'I sleep in my car. I eat in my car. I work in my car. That is not freedom. That is not flexibility.'"*

And that was only part of the wreckage. Over the next few years, hundreds of taxi drivers who had taken out large "balloon loans"—which typically have low interest rates but involve one large payment due upon maturity—to purchase medallions went bankrupt or teetered on the edge. A few died of the intense stress caused by loss of income and long hours. Others continued driving, but lost their housing, spending the night in their taxi or on the couches of friends. Some, desperate for work, began driving for Uber and Lyft, even as the companies were quickly slashing wages. These drivers, especially those for whom driving had been a lifetime profession or who owed money on a vehicle they had purchased for work, felt trapped. In one of the most expensive places in the world, drivers in San Francisco worked long hours with no control over their earnings. And within months, this legalized system of low pay, debt, and devastation was replicated the world over.

Beginning in 2016, without access to municipal regulators, fare controls, or vehicle caps, Uber and Lyft drivers turned to the only places left: the courts and each other. Like cabbies before them, ride-hailing drivers began to self-organize and form drivers groups. Some even worked with plaintiff-side employment attorneys to sue Uber and Lyft for millions in back wages and vehicle reimbursements. Uber alone faced dozens, and eventually hundreds, of lawsuits and complaints alleging that they were misclassifying their drivers as independent contractors.

But Uber and Lyft, even though they were unprofitable, had unprecedented sums of venture capital funding with which they could aggressively litigate such cases and confidentially settle them, sometimes for more than what the plaintiff would have received had they won at trial. The companies used these ungodly streams of capital along with legal acrobatics—both settlements and arbitration clauses—to avert a conclusive finding on the central issue: were drivers employees or independent contractors? By avoiding a legal decision at either the district or appellate level, the companies continued to get away with treating their workers as independent contractors.

Samy, an Iranian refugee who emigrated to California in the 2000s, described the terrible conditions that he and many of these "independent contractors" face. Unable to afford the Bay Area, he lives a hundred miles away from San Francisco, and rather than commute to the city daily, he sleeps in his car between shifts, sometimes not returning home for several days in a row. "We have no freedom," he told me. "I sleep in my car. I eat in my car. I work in my car. That is not freedom. That is not flexibility." Because he has to make a certain amount each day to make ends meet, Samy hypothesizes that Uber has used the data they have on him to systematically lower his wages, forcing him to work harder and longer to earn that same amount.

Back in Iran, Samy, who is a highly skilled carpenter, ran a small construction business. "In Iran, I had some guys, they were working for me... I took care of them," he said. "I was giving them breakfast, lunch, and I even paid for their dinner. In

America, it's something different than all over the world." He said he hadn't heard of another country where workers were required to have their own tools to work for a company, and then not considered employees on that basis. "It makes no sense."

Finding Friction

In the post-union, post-employment decades leading up to the appearance of Uber and Lyft in 2012, taxi driver advocates in San Francisco had largely avoided organizing around employment status. Because the issue was contentious among drivers, San Francisco taxi worker activists had relied mostly on policy advocacy to improve their working conditions, lobbying regulators to ensure the economic security of taxi drivers and their families. With the legalization of the Uber and Lyft business models in the mid-2010s, an industry of workers who had long understood themselves to be "independent" began to build and leverage collective labor power, not just policy advocacy, to demand changes to both the industry business model and the regulatory environment. Drivers of all kinds—full-time and part-time, Uber and Lyft and taxi—took to collective activity, including direct actions, which came to a head in 2019.

As Uber and Lyft approached their IPOs early that year, the companies noted in their public SEC filings that the classification of their drivers as independent contractors remained key to their future profitability. Uber and Lyft drivers quickly ascertained that their status was the companies' most painful pressure point. Moreover, the fact that the companies—whose executives made millions—were relying on avoiding paying minimum wage and overtime to the drivers for the companies' profitability enraged drivers. As one driver explained to me in the summer of 2019:

> *Four years ago, there is no way anyone would have [supported] employment. People were getting what they needed, and they felt loyalty to the company. But now we can't survive. And did you see that [Uber co-founder] Garrett Camp bought a seventy-one-million-dollar home in Beverly Hills? That is our money. That is some bullshit. I know people living in their cars. And he bought a seventy-one-million-dollar home. We made him every dollar he has.*

Until this moment, driver organizers had avoided the potentially divisive issue of employment status. But in response to the IPO, a changing legal landscape, and fury rooted in their extreme economic vulnerabilities, drivers adjusted their tactics to direct their energies at technocapital's Achilles' heel.

On May 8, 2019, a coalition of Uber, Lyft, and taxi drivers, led by the New York Taxi Workers Alliance and the California-based Rideshare Drivers United (RDU)—whose members had been self-organizing and protesting for better regulatory oversight of the industry for over two years—called the first ever global strike against Uber. In the weeks leading up to the strike, the demands focused on fair pay and an appeals process to deal with terminations, yet they conspicuously lacked the word "employment." But during the strike action in front of Uber's headquarters on Market Street in San Francisco, drivers were beginning to talk about AB5.

The bill, which had been introduced in the California State Assembly just a few months earlier, drew upon a recent state Supreme Court decision to change the test for employment status. Rather than examining how much control a putative employer exerted over workers, the decision called for the *presumption* of employment status. If a company wanted to use independent contractor labor, they had to pass an exacting "ABC test" which asked, among other things, if the workers were occupied in a line of work that differed from the company's core business. AB5 constituted a striking revision in California employment law and strongly favored extending basic protections like minimum wage, overtime compensation, workers' compensation, and unemployment insurance to gig economy workers.

Two days after the strike, Uber went public, and driver organizers with Rideshare Drivers United began to talk more publicly about AB5 and the possibility of employment status. If it passed, it would be the first time since Uber and Lyft's founding that drivers had a law on the books that could support their struggle against the companies. They decided, in spite of some fears, to throw their support behind the bill's passage.

Still, in private, drivers were ambivalent about formally becoming employees. "We get behind this law because we have to show the companies we have power," Soleiman, a full-time Lyft driver, told me privately, after an organizing meeting. "But what does it mean for me and other drivers?" Though worried about the consequences, he and other Uber and Lyft drivers had advocated loudly for AB5, putting their fears to one side and leveraging collective action to fight for the one thing that companies most wanted to avoid.

In a truly unexpected turn of political events, AB5 passed. The California legislature, which had previously shown great deference and affection for technology companies, sided with the workers on the issue of employment status. The Assembly's speaker, Anthony Rendon, went so far as to call the so-called gig economy "fucking feudalism" in announcing his own support of the law. Upon its passage, AB5 was celebrated by driver advocates the world over and signed into law by Governor Newsom in September 2019. It was the first sign in over eight years that, perhaps, worker power could fracture technocapital's tight grip over lawmakers.

But like Soleiman, many drivers did not fully understand the implications of the new law—not because they were ignorant of the statute, but because so much was left in the company's hands. What would it mean to be an "employee"? Most centrally, would the companies take away whatever degree of flexibility the drivers had? Could Uber and Lyft cap their potential earnings or push them off the app? Or would the law facilitate exactly what precarious workers in the economy needed: a wage floor, access to the safety net, and a pathway to unionization?

While debates about the impact and efficacy of AB5 and the ABC test roiled over social media and in courtrooms—Uber and the food delivery service Postmates sought and lost a preliminary injunction against the law, while simultaneously arguing that the law did not apply to them—workers privately debated what to do. The law was supposed to go into effect on January 1, 2020. The day came and went and, conspicuously, nothing changed. When it became clear to workers that neither the city attorneys nor the state attorney general was going to force Uber to end its practice of misclassification, drivers decided to take matters into their own hands.

"We cannot wait for the city or state to enforce this law," Nicole, a Lyft driver and organizer in California, said to me over the phone, weeks after the law had gone into effect. "We have to do it ourselves." On February 5, 2020, members and supporters of RDU marched together to the Labor Commission's offices all over California, collectively submitting individual claims for back wages and vehicle reimbursements against the companies. The roughly 150 drivers who filed claims against Uber and Lyft that day calculated that together they were owed up to $4 million dollars for the last three years of underpayment. AB5 was not an end in itself, RDU organizers explained to the press and to their fellow workers, but a route to collectively achieve much-needed security, a law they could use to stabilize income and aid driver organizing efforts.

> *"Employment status created a path through which workers could organize and maybe even get a federally-recognized union."*

"Look, the companies can threaten us all they want," Chris, a Bay Area Lyft driver and RDU organizer told his fellow workers at an organizing meeting. "They can keep sending us emails and these in-app messages. But what employment status means is we can organize together. We can build power. We can build a union. And then we can force them to bargain over our working conditions. And then they can't take away our flexibility if we won't let them take it away." For Chris and others, employment status created a path through which workers could organize and maybe even get a federally-recognized union. Talk of a union, which had been stymied for three decades in the taxi industry by the reality that taxi workers without a union contract could still use the municipal regulatory regime to eke out a living under the leasing apparatus, has become a live issue once again.

While taxi companies had pioneered independent contracting in an earlier moment, technocapital's flamboyant forms of neoliberal risk-shifting have catalyzed a powerful resistance to both the corporate practice of eschewing direct employment and worker reliance on state structures for security. The second neoliberal turn brought on by Uber and Lyft eradicated both municipal regulations and the limited sense of agency that workers had over their livelihoods. Ironically, this purge of state structures in the 2010s that drivers could leverage stirred new—but old—forms of collective activity. For the first time since the early 1980s, employment status—and more importantly, shared worker power built through the threat of employment status—has become a way for ride-hailing drivers to resist extreme economic insecurity and to collectively imagine better futures.

Workers are turning to one another to build power and effect change. Through protests, strikes, and other direct actions that have emerged through the fight for employment in California, drivers have found a much-needed source of friction that can stop, or at least slow, technocapital's assault on their lives: solidarity.

Veena Dubal is an Associate Professor of Law at the University of California, Hastings. Her research focuses on the intersection of law, technology, and precarious work.